Forensic DNA Transfer

Forensic DNA Transfer provides a guide to the recognition and current understanding of DNA transfer in forensic criminal investigations. Increased improvements in technology mean that it is now routinely possible to obtain DNA profiles from non-visible deposits. How or when the DNA in question was deposited may be an issue in the context of the case, especially if the donor of the DNA is not in dispute. A DNA profile alone cannot reveal when or how that DNA was deposited at a crime scene, nor can it reveal the body matter from which it originated. Issues of transfer associated with activities may be debated – which the traditional discrimination purpose of DNA profiling cannot address.

DNA may be everywhere and anywhere – in homes, at workplaces, during transport and on personal items including clothing. DNA from a person may be on an object they have never contacted or in a room they have never entered. Concepts discussed in the book include non-self DNA on hands through day-to-day activities, the prevalence of background DNA in the environment and perhaps on the exhibit, the persistence of any DNA transferred and that a DNA result will depend on these variables as well as recovery techniques. Since DNA may be transferred to an exhibit (a) during the commission of a crime, (b) before the crime and/or (c) after the crime through handling, examination and testing, this book covers various transfer pathways and sources of DNA. Inadvertent issues of transfer of DNA resulting in wrongful convictions and the misleading of investigations are discussed, with an emphasis on contamination mitigation.

Forensic DNA Transfer examines the additional complexity resulting from non-visible deposits of DNA that impact on sampling and testing regimes. The changing understanding of the composition of purported 'touch DNA' deposits from the skin, including extracellular DNA transported via body secretions is described. Further, the newer focus on interpreting DNA evidence – using activity-level propositions and the rationale and associated issues – is also discussed.

Jane Moira Taupin is an independent forensic science consultant and trainer. She earned her B.Sc.(Hons) in Chemistry and MA in Criminology from the University of Melbourne, Australia. She has attended crime scenes for biological evidence and presented DNA profiling evidence at court in multiple jurisdictions for over two decades. She has co-authored a book on the forensic examination of clothing and is the author of three books on DNA profiling in criminal cases for the legal and forensic communities. She has received several awards in recognition and support of her work from national and international sources.

Forensic DNA Transfer

Jane Moira Taupin

CRC Press
Taylor & Francis Group
Boca Raton London New York

CRC Press is an imprint of the
Taylor & Francis Group, an **informa** business

Front cover image: Iurii Motov/Shutterstock

First edition published 2024
by CRC Press
6000 Broken Sound Parkway NW, Suite 300, Boca Raton, FL 33487-2742

and by CRC Press
4 Park Square, Milton Park, Abingdon, Oxon, OX14 4RN

CRC Press is an imprint of Taylor & Francis Group, LLC

ISBN: 978-0-367-74206-5 (hbk)
ISBN: 978-0-367-74638-4 (pbk)
ISBN: 978-1-003-15884-4 (ebk)

DOI: 10.4324/9781003158844

Typeset in Minion
by Deanta Global Publishing Services, Chennai, India

Contents

Preface

Aims: This book provides a guide to recognizing and understanding DNA transfer in criminal investigations.

It has long been known that visible deposits of biological material containing DNA may readily have been transferred, and can subsequently be transferred, in a variety of ways. The ever-increasing improvements in technology have meant that it is possible to detect ever-smaller amounts of DNA from non-visible deposits. It is common practice for forensic laboratories to obtain a DNA result from a dozen human cells or less, and the type and number of crime exhibits submitted for DNA analysis in recent years has greatly expanded.

How or when the DNA in question was deposited may be an issue in the context of the case – a DNA profile alone cannot reveal when, or how, that DNA was deposited at a crime scene. Different DNA profiles, obtained from the same item, may have been deposited at different times and in different ways. DNA may be *transferred* to an exhibit (a) during the commission of a crime, (b) before the crime, and/or (c) after the crime through handling, examination and testing. This book will cover the principles and impact of DNA transfer in criminal investigations.

The principles of DNA transfer have evolved from the study of transfer of visible trace material such as fibres. When deposits of biological material are not visible, either with the naked eye or through microscopic techniques, physical characteristics that may be informative such as shape and size of the deposit are lacking. Any sampling and testing rationale is consequently impacted. A 'swab it and see' or speculative approach is used on non-visible deposits, which has accompanying issues that will be discussed in this book.

'Trace DNA' has replaced the term 'touch DNA' as it cannot be assumed that non-visible deposits of DNA originated from touching through skin surfaces. Trace DNA is defined as DNA from an unspecified body location. The inferred deposit may be swabbed, cut out or lifted by adhesive tape, and submitted for DNA profiling without any prior testing for somatic (body) location. The resulting consequences of increased uncertainty will be described. Further, the changing understanding of the composition of purported 'touch DNA' deposits from the skin of the hand (or other skin areas) – including extracellular DNA transported via sweat or sebum – is discussed. This understanding includes the realization that there may not be a fixed 'shedder status' for a particular person, and that a person's ability to deposit DNA via the skin exists on a continuum.

There has been an increasing fragmentation of forensic science services which has led to situations where the DNA reporting scientist may never have seen the exhibit or performed any testing at any stage. Rather, the reporting (and testifying) scientist may have reviewed the case notes and results of many others in the laboratory, and may not be aware of the collection and/or any preliminary examination and testing conditions prior to receipt at the laboratory. This 'silo' effect also includes any previous examination at other sites such as at the crime scene and at other laboratories, for example at a police laboratory where 'screening' of an exhibit is performed before forwarding a swab or a cutting from that item to a DNA laboratory. This book will discuss the problems this has created in documented criminal cases.

The current desire by the forensic community to perform 'activity' level analysis, where propositions address 'how' the DNA (especially trace DNA) may have been deposited, according to different activities, will be discussed. The concept of different activities as to how the DNA arrived in the deposit analyzed, is essentially changing the value of DNA evidence from its traditional value of discriminating between people. The knowledge base is growing, but studies to inform 'activity' level analysis have been criticized due to poor experimental design, and conflicting results have been produced. This book discusses the problems inherent in moving up the hierarchy of propositions, and why there is a recommendation to consider higher levels. When the DNA profile is the only DNA evidence in the case it has sometimes been given more weight than desirable, and miscarriages of justice and/or misleading police investigations have followed.

Research is addressing the transfer of trace amounts of DNA which includes persistence, prevalence and recovery of DNA in various experimental situations. Concepts such as non-self DNA on hands through day-to-day activities, substrate materials, the prevalence of background DNA in the environment and perhaps on the exhibit and that indirect transfer of material must always be considered, will be discussed. The research is continuing apace, and this book no doubt misses new valuable contributions by the time it is published.

Contamination issues have been highlighted in recent years regarding collection and analysis before receipt at the forensic laboratory, and from cold case investigations where the continuity is unclear or the exhibit was originally examined using procedures that pre-date the high sensitivity era of DNA profiling. This book will highlight contamination issues by describing current mitigation responses and illustration of the impact of contamination on documented criminal cases. Contamination issues may result from direct transfer (such as from an investigator), or indirect transfer from a multitude of ways.

This work aims to provide tools to assist the criminal justice professional when presented with a DNA result in a forensic report or testimony. The DNA result and any associated statistic may not reveal any ambiguity, complexity and/or the assumptions and limitations involved in deriving it. DNA transfer issues may need to be considered. Questions posed from resolved criminal cases involving transfer of DNA principles and the relevant forensic literature are provided to help the reader to assess a DNA result/s in the context of the case.

Scope: This book discusses the common DNA analytical methods used in criminal trials today; nuclear STR profiling techniques and Y-STR profiling. The literature quoted and discussed is by no means complete but provides a representative picture of the topics covered, until early 2023.

The nature of DNA transfer is discussed, beginning with basic concepts of trace evidence transfer. How the current literature addresses DNA transfer will be described with particular regard to DNA–TPPR; transfer, persistence, prevalence and recovery. A DNA result cannot be obtained unless the DNA has been transferred (T) in some manner from a person, and has persisted (P) on a surface with sufficient prevalence (P) to be able to be recovered (R).

The hierarchy of propositions is discussed as well as the use of the likelihood ratio and the formulation of alternative propositions. How to address the problems that still occur when conveying the meaning of the likelihood ratio and the understanding (and mis-understanding) of both forensic scientists and the criminal justice system is beyond the scope of this book, but the basic derivations are provided.

There is an emphasis on contamination issues, with the newer focus on the risk before receipt at the DNA laboratory. The risk of contamination in the DNA laboratory is also discussed, such as through evidence bags and case files.

Resolved criminal cases and the relevant forensic literature are provided for the reader to assess a DNA result and address issues regarding DNA transfer.

Specific forensic DNA transfer issues are illustrated with court proceedings such as trials and appeals, commissions of inquiry and government and laboratory reviews.

Case studies emphasize the need for conclusions in the forensic report and in testimony that are supported by the data.

Literature references to scientific papers are provided but it is not necessary for the reader to peruse any or all, but to note these for possible relevant cases. A forensic scientist providing DNA evidence at trial should be familiar with the literature. Trial lawyers will be assisted in their examination of the expert by quoting these papers and assessing the expert's knowledge of the matters discussed therein.

The following will be discussed.

- Principles of transfer of trace material; visible and non-visible.
- The 'silo' effect of the separation of DNA analysis from the crime scene, and now more often separation from the exhibit.
- Principles of DNA transfer as highlighted in the literature and high-profile criminal cases.
- Fundamentals of interpretation including the likelihood ratio and the hierarchy of propositions.
- Direct transfer and indirect transfer.
- Trace DNA versus 'touch' DNA; self and non-self DNA.
- The composition of so-called 'touch DNA' deposits.
- Background DNA.
- Sampling and testing rationales.
- Medical exhibit considerations – internal and external body swabs; includes Y-STR analysis.
- Personal items; clothing; instruments – knives and firearms; communal spaces.
- Quality procedures and documentation.
- Packaging, handling and storage.
- Staff elimination DNA databases.
- Contamination issues with the risk before receipt at the DNA laboratory, at the laboratory and re-examination of 'cold cases'.

The book is suitable for lawyers and attorneys, forensic scientists, other criminal justice professionals and students.

Appendices address the principles of DNA transfer (Appendix A), and flawed logic that may be used in the interpretation of DNA transfer (Appendix B).

A glossary of terms used in this book is provided.

Literature references are cited throughout the text.

Acknowledgements

I would like to thank my editor Mark Listewnik at Taylor and Francis for his interest and patience regarding this book.

I would also like to thank the many barristers and lawyers I have worked with who have devoted considerable time and effort in grappling with DNA transfer in their cases. Thanks especially to Scott Johns KC, Rahmin de Kretser, Kellie Blair, Ian Hill KC, Marie Shaw KC and Public Defender Elizabeth McLaughlin.

This book is for barristers, lawyers and attorneys who are confronted with a DNA report, and for forensic scientists confronted with an ever-changing landscape in trace DNA evidence. I hope you find it useful.

About the Author

Jane Moira Taupin obtained a Bachelor of Science (Honours) degree in Chemistry from the University of Melbourne in Australia. Her fourth-year research thesis was titled 'Fluorescent Probes in Polyelectrolyte Systems'. Upon graduating she accepted research positions at University of Melbourne research facilities. Firstly she worked in immunology and antibody production at the Howard Florey Institute. She then worked for 3 years on a project funded by the Anti-Cancer Council of Victoria on a drug used in the treatment of leukaemia, at the Austin Hospital in Victoria.

She joined the Australian Federal Police as a Constable and then Stage 1 Detective and worked in diverse areas including judicial protection, drug surveillance and government fraud in Sydney and Melbourne. During this time she was transferred temporarily to the only atomic energy facility in the country (Lucas Heights), using neutron activation analysis on a number of criminal cases. She left to join the Victoria Police Forensic Services Centre as a forensic scientist where she reported a wide variety of cases involving biological evidence in major crime, with her final position as a Premier Caseworker. Her work included attendance at crime scenes searching for biological fluids throughout the State of Victoria. She presented biological expert evidence in courts of law, and presented DNA profiling evidence in court since 1999. Concurrently she obtained a Post-Graduate Diploma in Criminology and then a Master of Arts in Criminology, both from the University of Melbourne. Her Masters' thesis in 1994 on the 'Impact of DNA Profiling' was one of the first in the field.

She then moved to Forensic Alliance in England where she performed similar work in their Oxford and Manchester laboratories, working on cases from police forces throughout England, Scotland and Wales. She performed blood stain pattern analysis at crime scenes throughout England. When LGC Forensics took over that company she joined as a lead scientist working on cases from police forces in the same regions as well as the Royal Military Police. In December 2009 she returned to Melbourne, Australia, and was employed at MRS Limited as an international forensic auditor and lectured in Qatar and Bahrain on a variety of subjects including DNA analysis.

She won a Young Investigators Award from the International Association of Forensic Sciences to attend their meeting in Tokyo in 1996 for her work on clothing damage analysis. The following year she won an Australian Government Michael Duffy travel fellowship to attend the American Academy of Forensic Sciences meeting in New York and international laboratories including the FBI, the Forensic Science Service in England and the BKA in Germany. She was invited and participated on the inaugural committee of SWGMAT (Scientific Working Group on Materials) under the auspices of the FBI in Washington D.C. for 6 years. In 2009 she was awarded a 'Good Citizen Award' from Greater Manchester Police in England for her work in helping to solve a horrific rape case on an elderly woman through DNA profiling evidence.

She is a member of the American Academy of Forensic Sciences. She was for 5 years a member of the DNA Consensus Body of this Academy Standards Board, producing standards for DNA analysis and interpretation.

Jane has published many articles in peer-reviewed journals on trace evidence, clothing damage and blood stain pattern analysis, as well as co-authored a book on the forensic examination of clothing. She has authored three books on DNA profiling in criminal cases for the legal and forensic communities published in 2013, 2016 and 2019.

Jane is currently an independent forensic consultant and trainer. She has reviewed cases involving DNA profiling and blood stain pattern analysis and clothing damage, predominantly requested by the defence, in Australia, the United Kingdom, the United States and Thailand. She has lectured to forensic scientists in Australia, the United Kingdom, Ireland, the United States, Qatar, Bahrain and Latin America.

Chapter 1

Transfer of Matter

BOX 1

- Principles of transfer.
- Transfer of visible trace material.
- Transfer of non-visible trace material.
- Transfer of visible and non-visible biological deposits.
- Technology enhancements.
- Changing forensic science landscape.

1.0 Introduction

This chapter introduces some principles of the transfer of trace material described in the literature since the early 1900s. Forensic DNA transfer uses these principles.

Trace material includes visible and non-visible deposits of material – composed of natural or synthetic, human or non-human matter. Human matter may contain cellular material with a core material of 'DNA', with a genetic code.

DOI: 10.4324/9781003158844-1

Deposits of human blood and semen have been traditionally examined to establish the association of forensic biological evidence and crimes of violence against the person. The advent of DNA profiling in the mid- to late 1980s revolutionized the discrimination power of the purported donor of these human deposits.

The meaning and success of DNA profiling depends on analyzing the relevant deposit and on the prevalence, persistence and recovery of the DNA-containing material – in context. In recent years the concept of 'transfer' in DNA analysis has been considered in a type of discipline called 'transfer, prevalence, persistence and recovery' abbreviated as 'DNA-TPPR' in DNA journal articles.

No forensic evidence can bring findings of certainty. There are limits to any scientific technique applied to crime scenes and exhibits – including DNA analysis. Criteria used to reason in the face of this uncertainty use probabilities and probabilistic reasoning. This probability depends on knowledge and information supplied to the examiner so that alternative explanations are considered. There are particular additional aspects to evaluate when assessing the value of DNA from substances that are not visible and at a cellular level compared to, for example, DNA associated with large quantities of blood.

1.1 Principles of Transfer

The principles of transfer of DNA-containing material are similar to that of any forensic *trace material* such as fibres, paint and glass. Locard's Theorem or Locard's Exchange Principle is the principle of trace evidence transfer. This principle was first enunciated by a French police scientist early in the 20th century (Locard, 1920). During a criminal act an offender will both leave trace evidence at the crime scene and take it away on their body or clothes. Note that the 'scientific method' is included in the title of the treatise (see reference) and is central to any forensic 'science' analysis; also that there may be an 'exchange' of material. The scientific method is discussed in Chapter 2.

Locard described the exchange of trace evidence as 'the silent witness'. He thought that every step, every contact, every touch leaves irrefutable evidence that is only diminished by human failure to understand and wholly utilize its value (Locard, 1920; see also Inman and Rudin, 2001, 2002 for principles). All potential traces should be considered, depending on the activity of the event and its information potential.

Sometimes Locard's theory is reduced to 'every contact leaves a trace', which does not consider the activities that produced the traces. Whether that trace is *detectable* depends on the quantity and quality of the material and analytical methods used in the detection. The *persistence* of the trace on items depends on the activities involving the trace, and the environment in which the activities occurred, after the trace was deposited.

Furthermore, as has been pointed out (Gill, 2014), the reduced theory is sometimes implied as 'every perpetrator leaves a trace'. Locard's Exchange Principle drives the expectation of the investigator that a DNA profile recovered from a crime scene must have something to do with the crime event. However it is not inevitable that a trace DNA profile will be recovered from the perpetrator. Donors of background DNA and investigator-mediated contaminants will automatically become suspects if the perpetrator is absent from the profile (ibid.). This is called the 'hidden perpetrator effect' by Gill (ibid.). Later chapters discuss these problems – also see Appendix B.

Direct transfer or primary transfer is the transfer of material from the source – such as a person or a piece of fabric or painted surface – to another object. Direct transfer of *biological* material includes dropping or coughing blood, spitting saliva, shedding hair, ejaculating semen and touching a surface.

Indirect transfer occurs when material deposited from an item or a person is then transferred (moved) to another person, item or surface. There has been no physical contact between the original depositor and the final surface on which the material is located. Any trace material including biological substances such as blood, semen, hair, saliva, urine, vomit and skin cells can be transferred in this manner. Further transfer may then follow. These steps are called secondary (for the transfer after primary), tertiary, quaternary, quinary and higher orders.

1.2 Transfer of Visible Trace Material

1.2.1 Early Experimental Data and Application to Crimes

Over 40 years ago the forensic literature described levels of fibre transfer. Generally this focused on *direct transfer* such as linking clothing from a suspect to the scene or clothing of a victim.

Secondary transfer was demonstrated in contacts between clothing and horizontal surfaces in 1975 (Pounds and Smalldon, 1975). Another paper (Lowrie and Jackson, 1994) investigated secondary fibre transfer via clothing and seats, specifically in movie theatres, and concluded that seats may be more important in this type of transfer in that they are more likely to retain a reservoir of fibres.

A 1987 paper reported that the secondary transfer of scalp hair is common during hair transfer even more so than direct transfers (Gaudette and Tessarolo, 1987). Alternatively, a study investigating pubic hair transfer during sexual intercourse found that even under optimum collection conditions pubic hair transfers from male to female and vice versa were observed only 17% of the time (Exline et al., 1998). These authors described direct transfer as occurring when a person transfers his or her own hair to an object, and secondary transfer of hair occurs when the hair transferred does not originate from the person transferring it. The authors described a paucity of research on the frequency of pubic hair transfer between individuals during sexual intercourse.

Understanding transfer of fibres gradually became focused on its role in contamination, especially when misinterpreted as evidence of direct contact between the accused and the crime scene. Contamination (or sometimes called 'pollution') is considered the addition of material subsequent to the offence or event and added during collection, handling, examination or analysis of the case.

Direct transfer through contamination by the examining forensic personnel a few decades ago was partly responsible for a miscarriage of justice in Canada. This resulted in a Royal Commission (Kaufman, 1998).

CASE 1.1 GUY PAUL MORIN AND FIBRE CONTAMINATION

Guy Paul Morin was charged in 1985 with the murder of his 9-year-old neighbour Christine Jessop. She had gone missing in October 1984 and her body was found in a field almost 3 months later, sexually assaulted and murdered. Morin was acquitted in the first trial. The second trial in 1992 convicted him, in part due to a link made between fibres found in his car and on her body. The Royal Commission of Inquiry found some alleged matching fibres originated through contamination from the examining laboratory, known to have been an issue in the particular case within the laboratory but not communicated to the court. It was possible that fibres from the jumper (sweater) of an examiner were 'dropped' onto the exhibit/s in question. Generally the laboratory examiners would work on up to 20 different cases at the same time, sometimes in the same environment, and laboratory coats were not changed between cases – if worn at all. There was a lack of documentation as to the personnel such as assistants that worked on the cases.

Limitations of the hair and fibre evidence were not explained in the initial trial of Morin. Only hair microscopy (DNA profiling was not routinely available) was used in the hair examination and the significance of findings of the hair evidence was greatly overstated. Similarly with fibre evidence including reference to an irrelevant study which supposedly bolstered the laboratory findings of direct transfer of fibres, and the lack of understanding of shared environments and fibres due to the proximity of the accused and deceased. Linked scenes of the case such as the body site and the car and residences were attended by the same investigators thus potentially compromising evidence collection.

DNA profiling of semen stains on the victim's underwear in 1995 revealed that Morin was not the donor, just days before his appeal. He was exonerated and later awarded 1.2 million Canadian dollars in compensation by the Ontario government.

During October 2020 it was announced that a 'DNA match' had been obtained from the semen stains from the victim's underwear. This was to Calvin Hoover, an associate of the victim's family who had died in 2015 (Johnston, 2021).

Case 1.1 highlights the importance of contamination mitigation in the laboratory regarding trace material and the limitation of any findings. The Royal Commission found that the contamination was a serious issue and ignoring such in the investigation revealed an *unconscious bias* in that the findings were only relevant to the case against Morin – and that the police now had to deal with contaminated evidence in their ongoing investigation (Kaufman, 1998).

The Commission findings also can be extended to the presence of DNA-containing material at crime scenes. The accused and the deceased lived in close proximity and had shared environments so the significance of fibres found in the case was greatly exaggerated. The significance of trace amounts of DNA should also be considered in DNA cases where people are in social or domestic situations.

Commissions of Inquiry such as the Kaufman Inquiry hold valuable information regarding the caution required and the limitations of practical application of a technique in the field and in the laboratory. Although the inquiry was held over two decades ago it is very relevant to all trace material, including DNA evidence.

1.2.2 Transfer of Debris

Transfer, persistence and detection are factors that affect findings of debris on clothing (Taupin and Cwiklik, 2010). Evaluation of transfer of visible debris in the examination of clothing can be conducted by performing a few shedding tests through examining the particular clothing in the case. However, association other than direct contact may include contact with other clothing worn by the same person, contact with another person in the same household or workplace, or with a site rather than a person (ibid.).

The transfer of domestic animal hair was a particular issue in another Canadian murder case involving 'Snowball', a cat belonging to the parents of an accused, although this time DNA profiling was used for both the blood stains and the cat hair (Menotti-Raymond et al., 1997). Many households have domestic companion pets such as cats and dogs. These animals may readily shed hair on the clothing and environment of humans, and these hairs are available for transfer.

CASE 1.2 'SNOWBALL' AND TRANSFER OF DOMESTIC CAT HAIR

The body of a 32-year-old woman Shirley Duguay was found in a shallow grave in a wooded area of Prince Edward Island in Canada in 1994, some 8 months after she disappeared. A man's leather jacket had been found 8 kilometres from her house 3 weeks after she had gone missing. The jacket was covered in blood stains with DNA profiles allegedly matching the deceased, but it also had several white cat hairs in the lining. The estranged partner of the deceased, Douglas Beamish, was living with his parents and they had a white cat called 'Snowball'. The cat hair on the jacket had a 'matching genomic DNA profile' to that of Snowball (Menotti-Raymond et al., 1997) and the case set a legal precedent allowing animal DNA to be admitted as evidence in criminal trials in the jurisdiction. Douglas Beamish was convicted of second-degree murder.

The deposit of cat hair in this case was corroborative evidence that linked the accused to the jacket, upon which the blood stains were recovered. The cat hair on the lining of the jacket of the accused could be considered 'indirect transfer' if the jacket had picked up the hairs from the living environment of the accused, or 'direct transfer' to the jacket if the accused had picked up the cat and the hair was subsequently transferred.

Fibre transfer had focused on direct transfer but during the 1970s to 1990s studies increasingly examined indirect transfer. A case published by this author (Taupin, 1996) illustrates the concept. A girl was allegedly abducted in a car and rape attempted by the accused driver who denied any contact. Clothing from both people and the covers from the front seats of the car were analyzed for trace evidence. Secondary transfer was indicated by dyed brown human head-type hairs (possibly originating from the wife of the accused) located on the car seat covers and the victim's clothing; the victim had un-dyed black hair. Secondary and possibly tertiary transfer was indicated by pink synthetic material comprising both small sections of fabric and fibres (possibly originating from the victim's mother/home who worked as a sewing machinist) located on the victim's clothing, a car seat cover and the clothing of the accused.

Secondary transfer was defined in the case as the first indirect transfer of fibres after primary or direct transfer, taking place via an intermediary object. Such secondary transfer may be followed by tertiary or higher transfer involving two or more intermediary objects.

The persistence of fibres has been denoted as a critical factor in the evaluation of fibre findings in forensic casework. A 2015 study examined fibre persistence on clothing that had been immersed in water (Lepot et al., 2015) which may be relevant to drowning or dumping bodies in water. The study found that the amount of protruding fibres and the density of the fibrous network of the surface of the recipient garment are key factors that increase persistence value (ibid.).

More recently debris research including fibres has focused on 'activity' levels – that is, evaluating the analytical findings of the debris such as fibres on the item in question with competing scenarios as to activities. The topic of activity levels is also being used in other trace material evaluation including DNA and will be discussed in later chapters.

A study examined contactless airborne transfer of textile fibres between different garments in enclosed spaces such as elevators in multi-storey buildings (Sheridan et al., 2020; Sheridan and Gallidabino, 2020). The study found that textile fibres are able to transfer between items in the absence of contact and both primary and secondary contactless airborne transfer can occur. Participants were in an elevator that operated as normal with people entering and exiting. The study noted that in a criminal case the finding of a few foreign incriminating fibres can be considered significant but in one experiment up to 66 fibres were transferred just by occupying an elevator. Variables noted such as composition of the garment and movement also impacted their findings (which of course is the situation with biological evidence). Finally the study noted that interpretation of fibre evidence at activity level requires extensive knowledge of all the possible transfer mechanisms that may explain the presence of fibres on a recipient surface.

1.3 Transfer of Non-visible Trace Material

The existence of tertiary transfer of gunshot residue (GSR), regarded as 'not visible to the naked eye' trace material, was confirmed (French and Morgan, 2015), and secondary transfer has been discussed with respect to different scenarios (Gassner et al., 2019). Transfer is characterized by three parameters: the source, the recipient and the environment (ibid.).

Gunshot particles may be deposited on the hands of an individual who is standing in close proximity of a discharge but who has not fired the weapon, by shaking hands with the shooter, and by handling the firearm after it has been discharged by another person. Interestingly, transfer was observed to the greatest extent in the study above when a police officer (who had fired a gun) arrested a person by handcuffing them on the ground, transferring original fired gunshot particles to the arrestee. The above authors state that alternative means of GSR deposition must be acknowledged when interpreting the presence of GSR on a surface or object. It shows how important secondary and tertiary transfer is when reconstructing firearm incidents. Further research is recommended to investigate GSR transfer in situations such as police arrests and transportation in a potentially contaminated police vehicle, including prevalence studies.

The principles of DNA transfer can be extrapolated from the transfer of visible debris in general. However, additional considerations are required if considering 'trace DNA'. This is because trace DNA may not be visible. Trace DNA is DNA that cannot be related to a specific body fluid or matter and may be composed of just a few cells. The examination of debris has been traditionally detected through its visual observation (whether by the naked eye or microscope). Furthermore, trace DNA can be transferred through unknown paths, whereas transfer of visible debris has traditionally been able to be followed through different, accepted, pathways.

1.4 Transfer of Visible and Non-visible Biological Deposits

Biological evidence – from which DNA derives – on a crime exhibit was originally analyzed due to a visible (through the naked eye or microscope) deposit, or through screening tests, potentially showing a link between the suspect, victim and/or the crime scene. This deposit was considered 'direct'; that is, there were no intermediary items. Blood from the victim was dripped onto the crime scene for example, or spurted onto the perpetrator, or the accused ejaculated semen inside the body of the victim. 'Contamination' or 'innocent transfer' arose if quantities of biological matter were added during collection or sampling of the exhibit, or perhaps innocently by a person rendering assistance. These matters were only recognized when other factors were considered in the case context.

How visible blood, semen or saliva was deposited has additional observational factors. An arterial blood spray pattern on a wall has a distinctive physical appearance (DiMeo and Taupin, 2012). A semen stain on the crotch area of a pair of underpants also has a visible notification to the examiner – perhaps using an alternative light source in the examination stage, or using a presumptive screening test. A saliva stain may appear as a 'dribble' on the front of a worn upper garment, again with the stain visualized with an alternate light source.

Even when a visible evidential sample can be related to a body fluid or matter, types of transfer should be considered – but may be complex in its interpretation. The understanding of the physics of fluid transfer in the interpretation of bloodstain pattern analysis has been recommended in an influential 2009 USA Government report (National Academy of Sciences, 2009).

A single DNA result may contradict all the other evidence in the case. A prosecution and trial in Australia used DNA not only to convict but as the only evidence that a crime was committed at all. A DNA profile from one medical item led to the 15-month imprisonment of a young man for rape in Melbourne, Australia – when in fact an offence never occurred (Vincent, 2010).

CASE 1.3 *R V. JAMA [2008]*

An unconscious woman was found in a locked toilet cubicle in a nightclub in 2006. She could not remember what had happened, and the police conveyed her to a hospital for a medical examination. A DNA profile was subsequently obtained in a forensic laboratory from one medical cervical swab taken from her by the examining doctor, and spermatozoa noted on the associated microscope slide/smear (1 intact spermatozoon and 15 spermatozoa heads). There were no other medical samples that indicated semen or spermatozoa.

A DNA profile from the medical cervical swab 'matched', via a database hit, the DNA profile of Farah Jama, a 19-year-old Somali origin male living with his family in Melbourne, Australia. It was concluded by the forensic scientist that it was 800 billion times more likely that the DNA would be observed if it originated from Jama than if it originated from another person. There was no other evidence.

Farah Jama was convicted at trial in 2008, sentenced to 6 years in jail and served 15 months in prison. This was despite him having an alibi from his family, saying he had never been to the suburb where the nightclub was, and not one person at the nightclub for more 'mature' patrons could remember seeing a young dark-skinned man that night. The conviction was overturned on appeal in December 2009. It was realized shortly before appeal that samples from an unrelated sexual incident involving Jama, where no charges were eventually laid, were taken by the same medical officer at the same location approximately 30 hours prior to medical samples from the alleged rape victim. Jama's DNA had been placed on the database as a result of the previous incident, and due to an apparent database 'hit' police obtained a reference sample from Jama and used it in the second case.

A State Government inquiry by a retired Supreme Court judge (The Vincent Inquiry) found that the DNA evidence had been the only link between Farah Jama and the unconscious woman in the toilet cubicle. Most likely, contamination between the evidentiary samples of the two cases occurred in the medical examination room, although the exact mechanism could not be determined – the possibility of transfer within the examining laboratory could not be excluded. The two cases had also been examined in the forensic laboratory by the same scientist, although at different benches. The inquiry stated:

> 'it is almost incredible that, in consequence of a minute particle, so small that it was invisible to the naked eye, being released into the environment and settling on a swab, slide or trolley surface, a chain of events could be started that

> culminated in the conviction of an individual for a crime that had never been committed by him or anyone else, created immense personal distress for many people and exposed a number of deficiencies in our criminal justice system. But that, I believe, is what happened'
>
> **(Vincent, 2010).**

The Vincent Inquiry could not ascertain the exact pathway of the transfer of the semen/DNA from the accused and reported that the deposition could have occurred in a variety of innocent ways. One way was that a dried semen particle was deposited on a swab and/or microscope slide. This could have happened as the patient from the first incident had dried semen in her head hair (deposited by ejaculation). The hair was cut during the medical examination to provide a sample for the forensic laboratory and the sample was subsequently analyzed in the laboratory. It was possible that flakes of that dried semen had been deposited in the environment of the medical examination room and on equipment such as the adjacent trolley, where swabs and microscope slides were left uncovered and exposed, instead of in a specially protected 'kit'. This author notes that examining the cut head hair for semen and a resulting DNA profile that was placed on the DNA database may also be an avenue for transfer of DNA if there was insufficient contamination mitigation as the same scientist examined both cases within days.

Modes of transfer postulated were 'indirect transfer' from the accused in the form of semen flaked from the hair of the first patient. This was agreed by the attorney of the accused as deposited via direct transfer from the accused to the first patient. The flakes may have landed on microscope slides/swabs/tools/surfaces used in her medical examination. The semen was then transferred to swabs/slides used in the examination of the second patient (indirect transfer). Regardless, this is a type of contamination or 'pollution' in the existing environment of the sampling process which has been described for many years in the literature and judicial inquiries.

Procedures in the jurisdiction were subsequently upgraded as a result of the inquiry. It is noted that as of writing the particular experiments required regarding the simulation scenarios of the Jama case have not been described in the literature. There may be ethical concerns preventing these types of experiments.

The above case shows the unexpected consequences when a deposit is not visible to the naked eye, even when its body origin can be ascertained. The inference in the original prosecution was that *transfer* of the semen of the accused had occurred through ejaculation into the vagina of the second patient, when in reality there was ejaculation into the hair of the first patient. Swabs and slides need to be DNA free before use in examination – equipment issues are further discussed in various chapters of this text. Further, the presence of semen on a swab from the complainant was implicitly related to the activity of sexual intercourse with the accused. This issue is discussed in further chapters.

1.5 Technology Enhancements and Limitations

1.5.1 Increasing Technology

The increasing sensitivity of DNA technology that allows tiny samples, invisible to the naked eye, to be analyzed for a DNA profile has meant that transfer principles are a necessary consideration in the interpretation of DNA evidence. Current analytical techniques are sensitive enough to produce informative DNA profiles from invisible deposits, even when no DNA has been quantified during testing.

When evidential DNA cannot be related to a designated biological fluid or matter such as blood, semen or saliva (known as 'trace DNA') there is greater uncertainty as to the method of deposition. Trace DNA cellular material is not visible, the particular somatic origin has not been tested – or been unsuccessful – and it could have been deposited on a surface by either direct or indirect (secondary or higher) transfer through a variety of ways.

DNA can be deposited by a person contacting an item but also can be readily transferred in various other ways including (a) from person to person, (b) from item to item, (c) from floor or surface to item, (d) from person to item and then another item, (e) from person to person and then item and so on. DNA from a person can end up on an item they have never touched or been near, or in a room or house to which they have never entered.

A biological substance that has been transferred multiple times, if detectable, may appear as components of complex mixture DNA profiles composed of many different contributors. This is because the vectors (such as hands or implements) aiding the transfer, and/or the substrate from which it is ultimately collected, may also carry DNA.

Items that have been potentially 'touched' have become important exhibits in all types of crime, from the most serious against the person to crimes against property. The increased sensitivity of techniques and the types of objects examined have also meant that many DNA profiles obtained are mixtures – DNA from multiple people together in the one profile. The potential scenarios in which the DNA may have been deposited have thus also increased.

A few years ago it was believed that, in addition to contact, DNA may transfer by aerosols (Fonnelop et al., 2015). Human cells already present in the surrounds, such as house dust, may be transferred to exhibits through their presence in the 'air'. Terms such as DNA 'falling from the ceiling' have been used in forensic laboratories (author information).

Recently it was described that both human and mammalian DNA can be detected in the air environment (Clare et al., 2021). This study investigated environmental DNA used as a tool for species biomonitoring and ecological research mainly in aquatic systems with a small amount of work in soil, snow and rain. The study demonstrated that DNA can be collected from air and used to identify mammals. The DNA sampled from air contained mixed templates which reflected the species known to be present within a confined space and that this material can be accessed using existing sampling methods (ibid.).

An even more recent forensic DNA study examined the feasibility of aerosol DNA as a source of information in criminal cases (Fantinato et al., 2022). A method to detect DNA in air was described and DNA was collected on air filters in rooms. These investigations were run in parallel with DNA collected from dust samples from undisturbed surfaces, such as ledges above doors. A full DNA profile was obtained from at least one of the occupants of the room from 13% of the air samples. Dust samples showed higher concentrations than air samples and resulted in complex DNA mixtures. Unknown DNA profiles were also found in air and it was postulated that there was subsequent secondary transfer from clothes, skin or personal items to the air (ibid.). The study also noted that the results 'will help to establish the propensity of detectable aerosolized human material from innocent individuals to spread at the scene' (ibid.).

A laboratory study examined the level of DNA that is deposited by an individual to their work environment (Puliatti et al., 2021). It was found that a person can deposit DNA in areas they were present, even if surfaces and/or objects were not contacted and even after only one day.

Another recent study found that 'contactless indirect DNA' may be transferred from fabric substrates (Thornbury et al., 2021). Agitation of clothing, pillowcases and towels found that DNA transfer frequently occurred and was possible from all three investigated items. The study authors state that there are many mechanisms and variables involved in how trace quantities of an individual's DNA may become located on a surface.

Our clothing, our home, our transport, our social contacts may all be sources of DNA. Certain articles may be a reservoir of cellular or extra-cellular matter, holding DNA from many deposits at many times. For example, bags, coats, hats, gloves and the insoles of shoes may be infrequently cleaned and contain DNA that has accumulated from the regular wearer and other contacts. This DNA may not only be from cells from the surface of the skin but also from other body secretions such as saliva, mucous, semen, nasal secretions and blood.

DNA is thus all around us and readily available for transfer in the environment of humans. It can be transferred from person to person or object to object, including investigator-mediated transfer at the crime scene. While the principles of trace evidence transfer have been known for many decades, transfer of trace DNA has only come into focus in recent years.

It is important to keep in mind that just because a DNA profile can be obtained from as little as a single cell, this does not mean that the source of the profile is relevant to the crime event under investigation (Gill, 2014). Without context, such evidence can be embellished and presented in a strongly biased way (Gill, 2019).

1.5.2 Context

1.5.2.1 DNA Profiling as a Discrimination Tool

DNA profiling is a powerful discrimination tool but the context of that DNA profile/s is paramount. DNA alone will not exonerate or incriminate. It is the interpretation of the DNA in the context of the case that has this potential, with limitations. The DNA evidence in a case may support the hypothesis or scenario proposed, or it may not. The evidence could indeed be equivocal. There are also numerous debates about the sole incriminating evidence of a DNA result in a case which is further discussed in this book.

A DNA result may be compromised through the scientific processes that yielded it or through the interpretation of its meaning in the context of the case (Taupin, 2013, 2016). An important element of the interpretation of a DNA result is the scientific method. Alternative explanations or scenarios should be considered by any scientific investigator in the evaluation of a particular result. Forensic analysis, including DNA analysis, should incorporate the scientific method at each stage, from the initial unveiling of the exhibit to its sampling and evaluation (Taupin and Cwiklik, 2010).

It should be noted that, still, research to evaluate the risk of indirect transfer has not kept pace with technological developments and the increase in sensitivity of DNA-typing techniques (Fonnelop et al., 2015).

The discriminating power of DNA profiling and resulting DNA profile from a deposit attempts to answer the question of 'who' may be the donor of the deposit. This is called the 'sub-source level' in the hierarchy of propositions. This hierarchy of propositions is discussed in later chapters, however sub-source is considered to be based solely on the DNA profile and discriminating power.

If the donor of the deposit is accepted by all then the justice system may be more interested in 'how' the DNA was deposited, which is called the 'activity level' in the hierarchy of propositions.

The discrimination power of DNA profiling cannot answer this question. Proper separation of these two levels is important – the justice system may apply the likelihood ratio assigned to the DNA profile to the alleged activity of how the DNA got there (Gill et al., 2020).

1.5.2.2 DNA Profiling as a Tool Comparing Competing 'Activities'

There are increasing efforts to evaluate the DNA evidence in terms of quantity and DNA profile composition in light of competing propositions as to 'how' and 'when' the DNA got to a deposit, called 'activity level reporting'. Advisory bodies particularly in the United Kingdom and Europe advocate the evaluation of results in light of competing activities (not just for DNA but other trace materials) when this is required to limit the risk of misleading the criminal justice system. It is needed when the amount of collected trace material is low and when considerations of transfer, persistence and recovery require specialized forensic knowledge. There is a widespread recognition that there is danger in leaving such assessments to non-scientists and it is their duty to guide the court (Biedermann et al., 2016).

Cases where the findings require an interpretation regarding 'activity level' are when the source of the trace material is not in dispute but how the trace material was transferred is debated (Taylor et al., 2018).

Assessment of findings at 'activity level' involves assumptions and calculations including many data variables based on controlled experimental data of similar scenarios and include concepts not only of transfer but of prevalence, persistence and recovery of the DNA results in question using mathematical tools such as Bayesian networks.

Such assessment is difficult to convey to the lay person and indeed the reporting forensic scientist. It is not correct to state that the DNA is more likely to have been deposited through direct transfer than indirect transfer (Taylor et al., 2018; Gill et al., 2020). This is commenting on the likelihood of one explanation of the results versus another. The scientist can only evaluate the value of the results obtained, in light of competing explanations as to how it was transferred through various enunciated activities. However, there are many documented cases where the scientist has stated in their opinion direct transfer was the most likely explanation for the results, which will be discussed in further chapters.

Statements such as 'secondary transfer was an unlikely explanation for the presence of the appellant's DNA on the door handle' as quoted from a court case in the United Kingdom by the DNA Commission (Gill et al., 2020) interweave the results with the propositions thus leading to the transposed conditional. The scientist is commenting on the likelihood of the scenario given the results, rather than the likelihood ratio of the results obtained given competing scenarios (although this text author notes that secondary transfer is not an explicit scenario). The complex nature of this subject is acknowledged by the DNA Commission in a footnote of the paper as 'previous courts of appeal in the UK have strongly denounced the transposed conditional, yet we see the fallacy here' (ibid.).

Far more research is required to understand the variables impacting DNA transfer and its associated elements of persistence, prevalence and recovery, referred to collectively as DNA-TPPR (van Oorschot et al., 2019). Indeed, the available literature on DNA transfer has been stated as lacking quality and systemization (Gosch and Courts, 2019). The extreme complexity of certain DNA transfer scenarios with numerous unknowns may rightfully mean the testifying scientist waive testimony as an objective answer cannot be given, thus following Wittgenstein's proposal that 'whereof one cannot speak, thereof one must be silent' (ibid.).

The topic is discussed in more detail in the following chapters.

1.6 Changing Forensic Science Landscape

1.6.1 Forensic Science Practice

The extension of technical frontiers should also be accompanied by conceptual developments and understanding (Biedermann et al., 2014). There should be an explanation of the assumptions, limitations and meaning in the context of the case in any report or communication of findings. The explanation should be in terms that an investigating officer and a court room without advanced scientific knowledge can understand.

The transformation of the forensic science laboratory into production line testing facilities is already under way (perhaps completed) due to the demand of DNA testing and automation. This is akin to clinical laboratories with tests by technicians with a predefined list of possible results (San Pietro et al., 2019). The rise of increasingly specialized scientists to the detriment of the generalist scientist (with knowledge of a number of techniques and a holistic approach) has led to the fear that non-scientist law enforcement personnel and attorneys are filling the void on the front- and back-ends of an investigation and litigation (ibid.).

This author has reviewed many cases where the reporting and testifying scientist has never seen the exhibit in question. During an Appeal Court hearing this author was handed an exhibit and the judges and the court gallery viewed the exhibit at the same time the reporting and testifying scientist for the prosecution viewed the exhibit for the very first time.

The reporting DNA scientist may collate a number of results obtained through a technical process by many people – and robots. Often, the exhibit has never been to the forensic science laboratory but has been swabbed, scraped or tape-lifted by police personnel for unconfirmed material at the crime scene or in a police laboratory.

The valuable step of the scientifically informed sampling and testing rationale by the forensic scientist (Taupin and Cwiklik, 2010) has sometimes been lost, minimized or compromised. A recent paper described that the role of the scientist in the investigation of crime has been increasingly confined to the laboratory and propose the discipline of 'traceology' to deal with the examination, analysis and scientific interpretation of event traces (signs or remnants) of earlier activities (Ristenblatt et al., 2022).

The above paper (ibid.) states that the era of the proactive, problem-defining, criminalist or generalist is over. Analysts are little more than reactive, protocol-constrained, laboratory technicians responding to routine requests from prosecutors and police. The absence of science at the front end of forensic investigations has resulted in biased, ineffective, inefficient and/or erroneous outcomes with immediate and long-term societal impacts.

Other problems have arisen in recent times due to the notion of a DNA result as the holy grail for any criminal case and the concentration of scientific endeavour in forensic biology laboratories on reducing the 'turnaround time' of a case – described as the time of submission of a sample to the issue of a DNA result.

DNA has been described as becoming 'fetishized' as the pinnacle of evidence, with evidential and statistical power pushed closer to its limits with smaller sample sizes striving to make the most of often low-quality forensic samples (Houck, 2019). Chasing new methods, straining existing ones, lacking validation of assumed methods may all result in missteps, accidents and errors (ibid.).

A recent inquiry into the operations of a government forensic DNA laboratory in Australia highlighted many issues, at the core of which was a concentration of quantity over quality. The particular laboratory 'has, for some time, focused on throughput and quick reporting of results to the detriment of high-quality science. That scourge has invaded many areas of the laboratory's practices: the validation of processes and equipment for use, the dedication of scientists' time to a proper review of cases, and the lack of resources for research, development and innovation' (Sofronoff, 2022).

Under the system that was investigated 'a scientist is not dealing with a case but with a "sample", the next item on the worklist, a product on a virtual conveyer belt … a whole of case review by a reporting scientist would only take place if all samples had to be collated into a witness statement for court. This happened in only about 10% of cases' (ibid.).

This text cannot address these concerns in detail but will note case studies and basic principles in which these problems are exemplified.

1.6.2 Forensic Science Literature

A USA government body discussing the forensic literature (National Commission on Forensic Science, 2015) noted that:

- it was unclear in some cases which literature citations were crucial to support the foundation of a particular forensic science discipline; and
- some of the cited literature had not undergone a rigorous peer-review process.

Concern has also been expressed regarding some biomedical studies in general. This was explained in a 'Science' journal editorial as, 'science advances on a foundation of trusted discoveries. Reproducing an experiment is one important approach that scientists use to gain confidence in their conclusions. Recently, the scientific community was shaken by reports that a troubling proportion of peer-reviewed preclinical studies are not reproducible' (McNutt, 2014).

A consensus on reporting principles to improve quality control in biomedical research and encourage public trust in science was published in the November 2014 *Journal of Nature* (Nature Editorial, 2014). It recommended that scientific journals should strongly encourage, as appropriate, that all materials used in the experiment be shared with those who wish to replicate the experiment.

Of course, the innovation of DNA profiling has led to the resolution of many (too great to quantify) cases that may never have been resolved or identified the potential perpetrator otherwise. This has been especially demonstrated in sexual offences with an unknown (to the victim) offender as predicted by this author in the mid-1990s (Taupin, 1994). Cases in which the offender is known to the victim and/or has a legitimate access to the crime scene have always been problematic even in the days when DNA profiling would only be attempted on visual blood stains or denoted seminal fluid from the female/male internal cavities.

More research is required in the forensic science disciplines. This is true even in DNA analyses considered to be the 'gold standard'. The interpretation of DNA evidence is much more than a statistical exercise and more research is required in the transfer, persistence and context of DNA found at a crime scene.

All scientific evidence is probabilistic. No current forensic technology can support unique identification of individuals or substances. DNA discrimination evidence between people is different only because it is explicit about its probability. Similarly, DNA transfer reasoning must be probabilistic and cannot be certain.

There have been attempts to introduce the judicial system to Bayesian reasoning especially with regard to 'activity' level propositions and trace amounts of material. These propositions consider issues such as transfer and persistence. This is essentially combining probabilities – see further discussion in this book regarding hierarchy of propositions. But the capacity of forensic experts to provide the guidance sought by the judiciary with respect to the transfer of DNA has been challenging and there is a need for further research (van Oorschot et al., 2021).

The following chapters describe the numerous factors that need to be taken into account when considering forensic DNA transfer in the context of the case.

References

Biedermann, A., Vuille, V., and Taroni, F., 2014, DNA, statistics and the law: A cross-disciplinary approach to forensic inference, *Frontiers in Genetics*, 5, 136.

Biedermann, A., Champod, C., Jackson, G., et al., 2016, Evaluation of forensic DNA traces when propositions of interest relate to activities: Analysis and discussion of recurrent concerns, *Frontiers in Genetics*, 7, 215–220.

Clare, E., Economou, C., Faulkes, C., et al., 2021, eDNAir: Proof of concept that animal DNA can be collected from air sampling, *Peer J*, 9e11030.

DiMeo, L. and Taupin, J., 2012, Arterial bloodstain patterns on clothing – An interesting case linking the accused to the scene, *Journal Bloodstain Pattern Analysis*, 28, 2, June, 3–10.

Exline, D., Smith, F., and Drexler, S., 1998, Frequency of pubic hair transfer during sexual intercourse, *Journal of Forensic Sciences*, 43, 3, 505–508.

Fantinato, C., Gill, P., and Fonnelop, A., 2022, Detection of human DNA in the air, *Forensic Science International: Genetics Supplement Series*, 8, 282–284.

Fonnelop, A., Egeland, T., and Gill, P., 2015, Secondary and subsequent DNA transfer during criminal investigation, *Forensic Science International: Genetics*, 17, 135–162.

French, J. and Morgan, R., 2015, An experimental investigation of the indirect transfer and deposition of gunshot residue: Further studies carried out with SEM-EDX analysis, *Forensic Science International*, 247, 14–17.

Gassner, A., Manganelli, M., Werner, D., et al., 2019, Secondary transfer of organic gun residues: Empirical data to assist the evaluation of three scenarios, *Science and Justice*, 59, 58–66.

Gaudette, B. and Tessarolo, A., 1987, Secondary transfer of human scalp hair, *Journal of Forensic Sciences*, 32, 5, 1241–1253.

Gill, P., 2014, *Misleading DNA evidence: Reasons for miscarriage of justice*, Academic Press Elsevier, London and New York.

Gill, P., 2019, DNA evidence and miscarriages of justice, *Forensic Science International*, 294, January, e1–e3.

Gill, P., Hicks, T., Butler, J., et al., 2020, DNA commission of the International society for forensic genetics: Assessing the value of forensic biological evidence – Guidelines highlighting the importance of propositions. Part II: Evaluation of biological traces considering activity level propositions, *Forensic Science International: Genetics*, 44, 1–13.

Gosch, A. and Courts, C., 2019, On DNA transfer: The lack and difficulty of systematic research and how to do it better, *Forensic Science International: Genetics*, 40, 24–36.

Houck, M., 2019, Tigers, black swans, and unicorns: The need for feedback and oversight, *Forensic Science International: Synergy*, 1, 79–82.

Inman, K. and Rudin, N., 2001, *Principles and practice of criminalistics: The profession of forensic science*, CRC Press, Boca Raton, Florida.

Inman, K. and Rudin, N., 2002, The origin of evidence, *Forensic Science International*, 126, 11–16.

Johnston, M., 2021, The hunt for a killer, *Toronto Life*, November 29 available at http://www.torontolife.com/city/inside-the-hunt-for-christine-jessops-real-killer/

Kaufman, Honorable F., 1998, *The commission on proceedings involving Guy Paul Morin*, Ministry of the Attorney General, Government of Ontario, Canada.

Lepot, L., Vanden Driessche, T., Gason, F., et al., 2015, Fibre persistence on immersed garment – Influence of knitted recipient fabrics, *Science and Justice*, 55, 4, 431–436

Locard, E., 1920, *L'enquête criminelle et les méthodes scientifiques*, Flammarion, Paris.

Lowrie, C. and Jackson, G., 1994, Secondary transfer of fibers, *Forensic Science International*, 64, 73–82.

Menotti-Raymond, M., David, V., and O'Brien, S., 1997, Pet cat hair implicates murder suspect, *Nature*, 386, 774.

McNutt, M., 2014, Editorial, Reproducibility, *Science*, 17, 3413.

National Academy of Sciences, 2009, *Strengthening forensic science in the United States: A path forward*, National Academies Press, Washington, DC.

National Commission on Forensic Science, 2015, *Scientific literature in support of forensic science and practice*, NIST, Department of Justice, USA Government, 21 January.

Nature editorial, 2014, *Journals unite for reproducibility*, 515, 7.

Pounds, C. and Smalldon, K., 1975, The transfer of fibers between clothing materials during simulated contacts and their persistence during wear, Part III – a preliminary investigation of the mechanisms involved, *Journal Forensic Science Society*, 15, 197–207.

Puliatti, L., Handt, O., and Taylor, D., 2021, The level of DNA an individual transfers to untouched items in their immediate surroundings, *Forensic Science International: Genetics*, 54, 102561

R v Jama {2008} VCC 0886, Victoria, Australia.

Ristenblatt, R., Hietpas, J., De Forest P., et al., 2022, Traceology, criminalistics, and forensic science, *Journal of Forensic Sciences*, 67, 28–32.

San Pietro, D., Kammrath, B., and De Forest, P., 2019, Is forensic science in danger of extinction? *Science and Justice*, 59, 199–202.

Sheridan, K. and Gallidabino, M., 2020, Forensic breakthrough helps explain how innocent people's clothing fibres could end up at crime scene, *The Conversation*, August 14.

Sheridan, K., Saltupyte, E., Palmer, R., et al., 2020, A study on contactless airborne transfer of textile fibres between garments in small compact semi-enclosed spaces, *Forensic Science International*, 315, 110432.

Sofronoff, W., 2022, *Commission of inquiry into forensic DNA testing in Queensland*, Final Report, Queensland, December 13.

Taupin, J.M., 1994, *The impact of DNA profiling on the criminal justice system*, University of Melbourne Master's Thesis, available at https://cat2.lib.unimelb.edu.au

Taupin, J.M., 1996. Hair and fiber transfer in an abduction case – Evidence from different levels of trace evidence transfer, *Journal of Forensic Sciences*, 41, 4, 697–699.

Taupin, J.M., 2013, *Introduction to forensic DNA profiling evidence for criminal justice professionals*, CRC Press, Boca Raton, Florida.

Taupin, J.M., 2016, *Using forensic DNA evidence at trial*, CRC Press, Boca Raton, Florida.

Taupin, J.M. and Cwiklik, C. 2010, *Scientific protocols for forensic examination of clothing*, CRC Press, Boca Raton, Florida.

Taylor, D., Biedermann, A., Hicks, T., et al., 2018, A template for constructing Bayesian networks in forensic biology cases when considering activity level propositions, *Forensic Science International: Genetics*, 33, 136–146.

Taylor, D., Kokshoorn, B., and Biedermann, A., 2018, Evaluation of forensic genetics findings given activity level propositions: A review, *Forensic Science International: Genetics*, 36, 34–49.

Thornbury, D., Goray, M., and van Oorschot, R. 2021, Transfer of DNA without contact from used clothing, pillowcases and towels by shaking agitation, *Science and Justice*, 61(6), 797–805.

van Oorschot, R., Szkuta, B., Meakin, G., et al., 2019, DNA transfer in forensic science: A review, *Forensic Science International: Genetics*, 38, 140–166.

van Oorschot, R., Meakin, G., Kokshoorn, B., et al., 2021, DNA transfer in forensic science: Recent progress towards meeting challenges, *Genes*, 12, 1–35.

Vincent, Honorable F., 2010, *Inquiry into the circumstances that led to the conviction of Mr. Farah Abdulkadir Jama*, Victorian Government Printer, Melbourne.

Chapter 2
DNA

BOX 2

- Meaning of DNA.
- Where DNA may be found.
- Discrimination of biological fluids/matter.
- Association of biological matter with DNA.
- The scientific method.
- The likelihood ratio.
- Hierarchy of propositions.
- Relevance of DNA evidence to criminal investigations.
- Relevance of DNA transfer to criminal investigations.

This chapter discusses DNA-containing material and why DNA transfer may be relevant to criminal investigations.

2.0 Introduction

DNA is the abbreviated form of 'Deoxyribonucleic acid'. It is a complex chemical and considered to be a genetic 'blueprint', responsible for our chemical and physical characteristics. The

DOI: 10.4324/9781003158844-2

genes carry information for making all the proteins required by the organism. These proteins determine how the organism looks, how well it fights infection and sometimes how it behaves. Other sequences of DNA have structural purposes or are involved in regulating the use of this genetic information.

Half our DNA is inherited from our mother (from the egg in fertilization) and half from our father (from the spermatozoa). Forensic DNA profiling examines locations along the DNA molecule that are highly variable from one individual to another. These regions were chosen because they were believed to have no known function in the human body and thus not come into conflict with laws regarding medical privacy. However, current research suggests that this may not now be completely true, as some regions chosen (for DNA databases for example) could be associated with expression levels for neighbouring genes that code for medical disease (Taylor, 2022).

Each person's DNA remains the same over their lifetime and the composition of the molecule remains the same throughout the body. This is a forensic advantage because the DNA of a bloodstain found at a crime scene can be compared with DNA from a reference saliva swab from a victim or a suspect.

The ultimate power of forensic DNA *profiling* is its power of discrimination as DNA is 'unique' to a person – only identical twins (or identical triplets etc.) will have the same DNA. One of the misconceptions that the public may carry is that DNA profiling examines all of the areas of the DNA molecule. This is impossible to do for every criminal case currently. In fact the entire 20,000 or more genes and the sequences of the 3 billion base pairs that make up the human genome in DNA had never been identified until the Human Genome Project completed a massive effort in 1993 (Human Genome Project). This original publicly funded project has continually improved efforts over the past 2 decades and the complete sequence of a human genome has now been published (Nurk et al., 2022).

It is even conceivable that in the future comparisons of whole genomes may be feasible and thus allow precise individual identification. At the moment, however, we are reliant on observation of a designated number of areas on the DNA molecule and statistical analysis in the comparison between the individual and the evidential sample. Furthermore, the DNA in each individual is approximately 99.9% the same as in another so that we are concerned with only very small differences in these sequences.

As well as preconceived ideas of the 'uniqueness' of DNA analysis, the recipient of forensic information may have the idea that the science is exact, precise and gives answers that are independent of human judgment (Hicks et al., 2022). This book endeavours to show that this is not the case.

2.1 Meaning of DNA

DNA is a stable molecule and does not degrade in non-hostile conditions. Exposure to harsh weather or bacterial attack may degrade the molecule (see Chapter 4 for a discussion on persistence). This stability is one of the reasons for replacement of traditional blood grouping and enzyme testing originally employed in forensic laboratories, as well as a better discrimination power. A rationale for the examination of 'cold cases' relies on the fact that while DNA analyses can be more discriminatory than previous blood-typing tests, DNA can also persist over time – although the amount of DNA detected is dependent on a variety of factors that are often unknown in a criminal case.

The advantage of DNA evidence over other forms of forensic evidence is that it enables a statistical approach based on genetic principles. The DNA testing kits have improved in specificity over the years so that more sites of the DNA molecule can be analyzed. This has increased the discrimination power of the tests so that more individuals can be excluded as donors.

However, it is generally not possible to DNA profile every person in a particular country – although increasingly there may be attempts. In the absence of whole population databases, sampling statistics are used to extrapolate frequencies of fragments ('alleles') used in the statistical evaluation of the DNA profile from a small database (sometimes less than a few hundred people) to a population at large.

The interpretation of DNA profiles when there is a so-called 'match' between profiles requires the determination of a probability; the chance of observing *that DNA profile* again, under various conditions and observed in a certain population. There has been much debate in the forensic literature about how this probability should be derived.

Advances in technology have enabled the forensic science laboratory to routinely detect enough DNA to generate a complete or partial DNA profile from just a few human cells, and sometimes from zero DNA quantified due to the greater sensitivity of the DNA profiling kits than quantification kits. Thus at the same time tests were more discriminatory they were becoming more sensitive and detecting more DNA. This DNA may or may not be relevant to the issue at hand, and from multiple contributors to the one DNA profile such that their relevance may be difficult to assess.

2.2 Where DNA Is Found

DNA resides in every cell of our body with the exception of red blood cells. The white cells in the blood carry DNA in their nucleus; mature red blood cells have no nucleus and thus no DNA. The term 'DNA' without a prefix generally implies nuclear (that is, in the nucleus of a cell) DNA.

Blood and semen have been the traditional fluids analyzed during a criminal investigation. Blood is associated with harm and rupture of the arteries and veins during an attack on the body. Semen is ejaculated by a male and is of importance in sexual assault investigations. Saliva may be associated with sexual offences. Due to advances in technology, DNA supposedly deposited through 'touching' through contact with the skin has been associated with all manner of offences.

The discrimination power of DNA profiling has increased the level of information that can be obtained from a crime scene in answering the question – Who did the biological material come from? When the persons involved in the alleged criminal activity are known to each other the identity of the donor of the biological material may not be in question. Rather, it is *how* did the fluid get there? This includes concepts of transfer.

An interesting case illustrating these principles is the complex case of David Camm in the United States that spanned over a decade with many twists. The case caused controversy not only in the media due to the horrific circumstances but also in the field of blood stain pattern analysis due to duelling expert opinions.

CASE 2.1 BIOLOGICAL FLUIDS AND THE DAVID CAMM CASE

One evening during 2000, a former police state trooper David Camm from Indiana returned home to find his wife and young son and daughter shot to death in their garage. His wife was found on the ground and his two children in a vehicle in the garage. Camm was charged with murder and bloodstain pattern experts in the original trial stated that the specks of blood detected on Camm's T-shirt were 'high velocity impact spatter' from shooting at close range (Gupta, 2017). Camm stated that the blood spots were produced through transfer from the hair of his daughter which was soaked in blood. The jury found him guilty of murder and he was sentenced to 195 years in prison. An Appeal Court overturned the ruling in 2004 finding that the jury had been prejudiced against him.

A year later the police reopened the case and, on request from the defence attorney, a DNA profile that had been obtained from the collar of a windcheater found underneath the body of Camm's son (foreign to the scene) was submitted to the national DNA database. This DNA apparently could be associated with Charles Boney, a person who had recently served time for armed robbery. The windcheater was prison issue and on the inside collar were the words 'BACK BONE'. The inside collar had been sampled for DNA at the start of the investigation but the resulting DNA profile had not previously been entered into the national DNA database.

Boney accepted that the windcheater was his but stated that he had sold a gun to David Camm wrapped in the windcheater, and that Camm had killed his family. Boney was sentenced to 225 years in prison as an accomplice to the murders. Apparently there were also deposits of DNA that could be attributed to Boney on the body of Camm's wife.

Camm's second trial for murder in 2006 featured conflicting scenarios as to the deposit of the blood stains on his T-shirt. The defence experts stated that the bloodstain patterns were 'transfer' stains produced as Camm described. The jury again found Camm guilty and he was sentenced to life in prison without parole.

> *There was an appeal in 2009 due to prejudicial conduct in the second trial and a third trial occurred in 2013. The bloodstains on the T-shirt were again debated as to deposition (essentially transfer mechanisms). Another expert stated that there was too little information in the patterns on the T-shirt to draw a conclusion as to the method of deposition, whether by direct deposit through a gunshot or through transfer by contact with head hair coated in blood (Schneider, 2013). Camm was found not guilty after spending 13 years in prison.*

Bloodstain pattern analysis (BPA) is a type of pattern interpretation technique. It is the study of the mechanisms of blood as a fluid when shed from the body and the patterns produced during the shedding or subsequent actions on that blood fluid deposit, such as smearing. However, the theoretical concepts that govern the evaluation and interpretation of bloodstain patterns on smooth, static non-porous surfaces found at crime scenes cannot be used to underpin the assessment and interpretation of bloodstains deposited on fabrics. The variability of textile surfaces – such as fabric construction, composition, treatment and surface curvature – will impact on the physical characteristic of a stain (Reynolds and Silenieks, 2016; Taupin and Cwiklik, 2010).

A report published by the National Academy of Sciences (2009) entitled 'Strengthening Forensic Science in the United States: A Path Forward' singled out many forensic science disciplines, including Bloodstain Pattern Analysis, for critical review. The acknowledgements that 'faulty forensic science has, on occasion, contributed to the wrongful conviction of innocent persons' and that many forensic techniques have never been subjected to stringent scientific scrutiny provided the impetus for the NAS's report, highlighting current limitations of forensic techniques. BPA was one of several disciplines that were criticized and it was stated that 'uncertainties associated with bloodstain pattern analysis are enormous'. Criticisms were made of the perceived over-reliance on an expert's personal experience given 'the importance of rigorous and objective hypothesis testing and the complex nature of fluid dynamics' with the opinions of experts held as more subjective than scientific (ibid.).

The Camm case above relied on observations of the appearance of blood stains and how they may have been deposited or transferred where the persons involved were known to each other. The DNA on the windcheater was accepted by all parties as 'wearer DNA' and Boney's windcheater, but his defence team claimed that this did not put him at the scene and an explanation was provided. Note that the physical characteristics of a stain cannot be applied to trace DNA including 'wearer DNA'.

Analysis of potentially 'touched' objects (worn clothes or handled items) for DNA is now a significant part of laboratory casework. Some jurisdictions have recorded that more than half the number of samples processed for DNA profiling were those potentially 'touched' (Baeschler, 2016). This author clarifies the meaning of 'touched' objects in Chapter 3.

Exhibits that involve serious crime with visible deposits, related to blood or semen, have been traditionally submitted for DNA analysis since the late 1980s to early 1990s in westernized countries. There is thus a greater knowledge base regarding visible biological deposits of identified material compared to exhibits potentially holding 'trace DNA'. Blood and semen are associated with violent or intimate contacts. However, trace DNA from an unspecified cellular source reduces the relevance of such biological evidence to a crime.

The first case/s where DNA profiling was used (initially called 'DNA fingerprinting') was a serial rape and murder case where quantities of semen with spermatozoa were found in the vaginas of both deceased girls (Gill and Werrett, 1987; Wambaugh, 1989).

CASE 2.2 THE PITCHFORK MURDERS

A 15-year-old girl Lynda Mann was found raped and murdered in 1983, abandoned in the English midlands countryside. Three years later, 15-year-old Dawn Ashworth was raped and murdered nearby, again abandoned in woodlands. Police asked geneticist Professor Alec Jeffreys of the University of Leicester to analyze samples from the vaginas of both girls using his then new technique of 'DNA fingerprinting'. The semen from both bodies had the same DNA profile. This led the police to conduct a world first DNA-led intelligence screen of over 5000 local men, with villagers providing blood samples in a mass testing. This was at the time when reference samples for DNA profiling were collected using blood from the individual and not the saliva samples used today.

A local baker, Colin Pitchfork, persuaded a work colleague to donate a blood sample for him. Police discovered this ruse and it was subsequently found that the DNA profile obtained from the semen on the bodies matched Pitchfork's DNA profile.

The 'source' level of semen was inferred for the DNA profile from the vaginal swabs. The 'activity' of vaginal sexual intercourse by the unknown offender through transfer of his semen was inferred.

The amount of DNA borne per volume of sample material or exhibit varies according to the source. Solid human tissue and sperm samples contain very large amounts of DNA per unit volume. Blood has the next highest DNA potential. Saliva and nose and mouth secretions exhibit the third-highest DNA potential because of the small volume of body fluid conveying DNA-bearing cells and small contact areas. 'Wearer' DNA is a type of trace DNA proposed to be found on clothing generally from contact with the skin of the wearer of the garment but the quantity depends on the DNA deposited and the time elapsed. A garment may act like a reservoir of DNA if washed infrequently (such as coats, hats and bags). Trace DNA in general, such as that proposed in handling scenarios, due to its transient nature has the lowest DNA potential.

Figure 2.1 lists the relative DNA contents of biological fluid or material.

Type of biological matter	Cell type containing DNA
Human tissue	Multiple including blood cells, epithelial cells, nerve cells
Semen	Sperm and epithelial cells
Blood	White blood cells
Saliva and nasal secretions	Epithelial cells, white blood cells
Vaginal secretions	Epithelial cells
'Wearer' DNA	Epithelial cells
Skin	Nucleated and non-nucleated epithelial cells, extra-cellular DNA

DNA *decreasing* in quantity

FIGURE 2.1 Sources of DNA. Note: Blood cells may exist from all types of matter due to proximity of veins and capillaries to the skin, and may be associated with harm (for example blood nose with nasal secretions). Blood may exist in vaginal secretions due to menstruation.

2.3 Association of Biological Matter to a DNA Profile

2.3.1 Body or Somatic Origin

A DNA profile does not reveal what body fluid or matter produced the DNA.

Previously, reference samples for comparison purposes from persons of interest were obtained from blood. The DNA was obtained from the nucleated white blood cells. Due to technology improvements reference samples are now taken from inside the mouth, which is less intrusive. It is accepted that most of this DNA is from the inside lining of the cheeks and are epithelial cells. These types of cells are those found lining the inside of organs and are oval shaped with a nucleus containing DNA.

Confirmation of blood and semen uses serology tests – to detect the presence of haemoglobin found in red blood cells or spermatozoa found in semen. These serological tests have been adapted from clinical testing for use in medical examinations and research.

It is the white blood cells that are nucleated and contain DNA (not red blood cells). Searching for the presence of spermatozoa (containing DNA) in semen involves staining of an extracted crime scene sample on a microscope slide and using a microscope to visually search for spermatozoa.

Traditional identification methods for the nature of the body fluids or tissues present encompass a narrow range and require visible material for screening (sometimes called presumptive) and then confirmatory tests. When there is little material available, the examiner may choose to sacrifice the preliminary test and rely on the confirmatory test. Presumptive tests for a biological fluid, by their very definition, are not confirmatory tests. If it is required to confirm blood, semen or saliva in a sample, confirmatory tests should be performed.

Optimal sampling may be compromised when there is no visible matter and the examiner resorts to speculative sampling of an area to obtain a DNA profile. The end result is that the examiner can only denote 'trace DNA' if a DNA profile is obtained.

Current forensic laboratory tests cannot show that a DNA deposit is from skin cells. It also cannot be assumed that touched objects contain only low amounts of DNA. Objects that are simply touched may contain quantities that may be encountered from blood or semen stains. It is the quality of the DNA that is important. Blood or semen deposits may be degraded or inhibited and produce poor quality DNA profiles.

2.3.2 Source Attribution

Denoting a deposit as having derived from a particular body fluid from a confirmatory test such as for blood may be a reasonable assumption if (a) the case context suggests such, (b) the scientific method is used and (c) a single source DNA profile is obtained. However, the problems with using presumptive (screening tests) and internal laboratory tests that were not internally or even foundationally validated was graphically illustrated nearly 30 years ago in the 'dingo' case from Australia.

CASE 2.3 THE CHAMBERLAIN CASE

Baby Azaria disappeared from a campsite near Uluru (Ayers Rock) in the desert of central Australia in 1980. Lindy Chamberlain said that a dingo (a wild dog) took her baby from their tent. She was convicted at trial of murdering her baby inside her car; there was supposed arterial blood staining under the dashboard where it was alleged she had cut the baby's throat. Lindy was sentenced to life imprisonment. Her husband was convicted of being an accessory to murder and given a suspended prison sentence. The child was never found.

A few years later there was an Australian Royal Commission that enquired into these convictions (Morling, 1987). Presumptive testing for blood (orthotoluidine) inside the car used a test that also cross-reacted with the minerals found in the red dirt and dust the car had travelled through in the desert. The Commission found that the laboratory test used to denote that there was foetal blood from a baby inside a car was not specific, and gave a positive result to 'sound deadener' used as a coating inside the car. The convictions were vacated. Lindy Chamberlain had served 4 years in prison. A final coronial inquest in 2012 found that, indeed, a dingo had taken Lindy Chamberlain's baby (Morris, 2012). During the intervening years there had been numerous attacks by dingoes on children.

Preserving positive results for biological fluid confirmatory tests, by either photographing or retaining dried samples of gels/tests, is necessary for independent review. Witnessing the result of a confirmatory test is a usual laboratory practice but does not reveal if there are systemic problems in the laboratory.

Presumptive or screening tests are generally not photographed or preserved, or even witnessed by another scientist. Sometimes these tests are performed at crime scenes by police officers. If there is sufficient visible material to perform a presumptive/screening test, then there should be enough material to perform a confirmatory test. Presumptive tests may be suitable for investigative purposes but should not be used for evidentiary purposes or in reports to the court.

Screening tests used to locate potential deposits that are not visible are used when large areas are required to be examined – such as bedding and clothing in sexual assault cases for semen stains using the seminal acid phosphatase (AP) test. Another example is the 'luminol' test that searches for potential blood stains that may have been diluted so much that they are no longer visible to the naked eye. These tests rely on the activity of substances in the fluids (not the DNA) to produce a colour change indicating it may be worthwhile to perform further testing for source origin once a positive area is located. Because these tests rely on the activity of the substances they are non-specific and other substances may produce positive reactions. Confirmatory tests analyze for the presence of the substance using antigen-antibody tests, or differential staining for the presence of spermatozoa under the microscope.

Even the so-called confirmatory tests such as those testing for saliva (RSID saliva) or those for prostate specific antigen (P30) used in testing for semen, have false positives or cross-react with known substances. A 'positive' result for an RSID saliva test on vaginal swabs does not mean that saliva is confirmed and many laboratories do not use this test for vaginal samples on the crotch of underpants for female complainants due to the known positive reactions from vaginal secretions (Sari et al., 2020). (See Chapter 5 for further discussion regarding sexual offence exhibits.)

It is necessary to distinguish between screening tests and the traditional confirmatory tests for blood and semen. Screening tests (sometimes called presumptive) use the 'activity' of a substance in a biological fluid to denote a positive result, whereas confirmatory tests aim to detect the presence of the substance. In this author's experience the traditional confirmatory tests have lost favour due to the additional work required, the ease of screening tests at the crime scene and the limited sample often analyzed that leaves none remaining for further testing. Now there is an argument regarding confirmatory tests not being 'confirmatory' at all (Hicks et al., 2022; Meakin et al., 2021). There have always been false positives in confirmatory tests going back to the 1980s, notably for human blood such as the 'double-immunodiffusion Ouchterlony test' recommended in the Chamberlain Royal Commission. These false positives were notably minimal and irrelevant – positive reactions were found for higher primates and ferrets, which should not be an issue in most crime cases. This is in contrast to the RSID saliva and P30 tests used today, which may react with other body fluids and is a serious issue.

The problem is that no test is 100% certain in a specific situation as laboratory errors can occur and no test is perfect, and thus it is necessary to state any assumption and limitation of a test used in the body of the scientific report. However, in this author's opinion, the most discriminatory test available should be used in an exhibit examination. If there is minimal deposit available to test, this should be reserved for scientific analysis at the laboratory. Scraping, cutting, swabbing samples or otherwise consuming and changing deposits for testing at the crime scene or in the police crime laboratory in an endeavour to get a positive result may compromise any further analysis in the forensic biology laboratory.

A *mixture* of body fluids may exist in the one stain producing a mixture DNA profile. A particular body fluid, or mixtures of such, cannot be determined from a DNA profile alone. Mixture DNA profiles may have a major contribution from cellular material and a minor contribution from blood, or vice versa.

Staged approaches are recommended to collect samples if mixtures of matter are predicted. For example, analyzing material underneath fingernails from a victim where foreign material is suspected – and there may be blood – requires consideration of mixture sampling. An attempt to minimize donor DNA and maximize foreign DNA can be employed. Blood staining may originate from the fingernail donor themselves due to the blood-letting that resulted from the event.

If the deposit of interest contains a good quality and large quantity of DNA, such as a fresh visible blood stain, then this DNA may overwhelm any 'background DNA' already present on the item. However, if the deposit of interest is trace DNA from a non-visible stain, the DNA from the deposit may be of a similar order, or in some cases may be overwhelmed by background DNA.

This aspect was first noted in a United Kingdom study (Peel and Gill, 2004). The study found that when dilute blood/saliva samples were deposited on a glass slide that had been 'handled' before or after the deposit, DNA could be recovered from both the body fluid and the handler. The respective ratios were dependent on the volume and concentration of the body fluid and the 'shedder status' of the handler. This is why mixture DNA profiles from surfaces may be especially problematic. Control samples from adjacent areas of the surface may be one solution (see the following chapters for sampling rationales and background DNA).

2.3.3 Association Error

The 'association fallacy' or error occurs when the strength of the evidence of the DNA profile is wrongly transposed to include a defined body fluid such as blood or semen in the calculation (Aitken et al., 2010; Gill, 2014). Observing body fluids, and detecting DNA profiles, use two separate testing methods. The fallacy also involves wrongful association of the presence of a body fluid such as blood with the activity in question, or even the probability of guilt (see also Section 2.7 regarding hierarchy of propositions).

Care is needed to explain to a court that the mere presence of a DNA profile does not automatically imply an associated 'activity'. The Jama Case (Case 1.3) is an example of a miscarriage of justice because the value of the DNA results was demonstrably carried over from sub-source to offence level propositions (Gill et al., 2020).

An association error occurs, for example, when relating the DNA profile probability statistic to the biological matter proposed. It happens when referring to extremely strong support that the accused is the source of blood from various samples. The statistical weighting should be based on the DNA profiles obtained compared to reference DNA profiles, and cannot be used for the association to a biological matter such as blood.

The association error may also occur when the forensic scientist states in testimony or reports 'in my opinion, the DNA came from semen' (or other biological fluid). Even worse, material denoted as 'possible blood' or 'possible saliva' using screening tests are stated in reports, which is not an evaluation at all but an investigatory avenue. This aspect is further explored in the following chapters.

2.3.4 Transfer Association Error

Recently there have been further association errors committed associating the strength of the likelihood ratio and quantity and quality of DNA to the method of transfer, such as direct or indirect transfer. It is a fallacy to state 'in my opinion, direct transfer is more likely than secondary transfer' which this author has noted in cases reviewed, cases highlighted in the media, cases quoted in courts of appeal, and even forensic DNA papers. When answering questions in court such as 'is it possible that there was secondary transfer?' one would be giving an opinion on the alleged activities implying secondary transfer (Hicks et al., 2022).

Experts may also be asked on the witness stand to agree with the proposition that direct transfer is more likely in a particular situation than secondary transfer, as described in a study on DNA transfer in an operational forensic laboratory (Taylor et al., 2016). An initially deposited amount of DNA will be lost with each transfer step in a chain and a seemingly logical extension might suggest that most of the DNA we detect is from primary transfer events. This thinking, however, does not take into account the number of potential pathways for these transfers to occur. It is in fact transposing evidence and propositions (ibid.). It is accepted that when obtaining a DNA profile then Pr (DNA/primary transfer) > Pr (DNA/secondary transfer). However, given the number of higher-order transfers that may occur regarding a transfer event it is likely that Pr (secondary transfer) > Pr (direct transfer). But courts often want to know Pr (direct transfer/DNA) versus Pr (secondary transfer/DNA) which is transposition of the conditional and not able to be determined by scientific means (ibid.).

One article has stated that scientists who object to 'activity level' propositions seem to have no objection to be asked, and to give, opinions in court about the probabilities of various competing activities regardless of whether relevant data exist (Biedermann et al., 2016). Analysts are likely to express themselves non-numerically in terms of transfer such as 'that is improbable', 'I don't believe that is likely to have happened', 'I think that sort of transfer is barely feasible'. Thus analysts who would be willing to express themselves in such probabilistic terms about propositions but refuse to provide probabilities for findings given propositions show an inherently contradictory position (ibid.).

Other statements of opinion such as 'The recovered DNA is the result of primary transfer' versus 'The DNA is the result of secondary transfer' are explanations, not propositions. They confuse the phenomenon of transfer (a variable conditioning the evaluation) with the alternative activities of interest (Taylor et al., 2018a). Activity-level propositions (see below) must specify alleged activities by a person. Transfer is an event of interest, about which uncertainty exists, and that is and should be taken into account in the scientist's assessment when evaluating findings given activity-level propositions but it does not define competing propositions in the first place (ibid.).

Statements using 'activity level' propositions aim to help address issues of indirect transfer versus direct transfer and the time of the activity, but it is important to avoid use of the word 'transfer' in propositions (Gill et al., 2020). (See Recommendation 3: 'Scientists must not give their opinion on "what is the most likely way of transfer" (direct or indirect) as this would amount to giving an opinion on the activities and result in a prosecutor's fallacy' (ibid.)).

One study devised questions to forensic science participants such as, 'assess whether they [the DNA profiles] originated either due to primary or secondary transfer, and indicate the likelihood of their chosen answer (definitively primary or secondary transfer; most likely' (van Oorschot et al., 2017). In this study, the scientists are thus not testifying about the probability of the evidence given the proposition. Instead, they are expressing opinions on events of transfer that the recipient of expert information may wrongly interpret as conclusions about competing activity-level propositions (Hicks et al., 2022; Taylor et al., 2018a).

This study also revealed disturbing findings as differences were observed between individuals from between and within laboratories on their understanding of DNA transfer, persistence, prevalence and recovery (DNA-TPPR). Self-perceived experience and expertise did not necessarily correlate to an increased number of correct answers, with several instances of poor performance from highly confident experts. The provision of additional scenario information did not appear to increase correct response rate either (van Oorschot et al., 2017).

2.4 The Scientific Method

Science is a method of study used to understand and describe the physical universe around us. The discipline of science is defined by the notion of hypothesis testing. First a hypothesis, or theory, is proposed. Experiments are then performed to test this hypothesis. The results of the experiments will either support or refute the hypothesis. The scientific method provides the framework for the testing of the hypothesis. There should always be alternative hypotheses provided (Inman and Rudin, 2001; Taupin and Cwiklik, 2010).

Appropriate data in support of a conclusion should be made available in the form of publicly available (and published) validation studies. The forensic scientist should be conversant with relevant specialist literature *including criticism*.

The scientific method incorporates reproducibility, rigour, transparency and independent verification. This is all part of a quality system that any reporting forensic laboratory requires.

2.5 The Likelihood Ratio

2.5.1 Origin

The likelihood ratio is derived from Bayes' Theorem concerning probability. Bayes' Theorem is named after the Scottish mathematician Thomas Bayes who lived in the early 18th century. He first used conditional probability to provide an algorithm statistically updating the *prior* probability of an event to a *posterior* probability of an event by the evidence observed in the matter, essentially the likelihood ratio. That is, posterior odds = *likelihood ratio* of observed evidence x prior odds. Or in formulaic terms:

$$\text{Prior odds} \times \text{LR} = \text{posterior odds}$$

The particular physical evidence considered in the case is not the only evidence used to determine the cause of an event and so cannot equate to the probability of the event. This consideration is important to remember when thinking about the meaning of the likelihood ratio and conditional probabilities upon which it is based.

An important point to remember is that if the prior odds are zero then the posterior odds are also zero – the likelihood ratio is irrelevant. This can be seen in the case of Lukis Anderson, who was charged with murder and had a high likelihood ratio of the DNA evidence, yet was in hospital at the time of the murder (see Case 4.1).

Statistical calculations using likelihood ratios can potentially cope with ambiguous profiles including artefacts produced as a consequence of low levels of DNA or stutter in the copying process, treating artefacts in a probabilistic manner. Thus the likelihood ratio is suitable for trace DNA evidence where the quantity and quality of any DNA profile produced cannot be predicted, unlike a pool of blood.

According to the recommendations of the DNA Commission over many years the evaluation of the DNA findings should be reported using a likelihood ratio (LR) approach based on case-specific propositions (Gill et al., 2006, 2012, 2018, 2020). The scientist assesses the value of the evidence in the light of two alternate propositions.

2.5.2 Formulation

A likelihood ratio involves the calculation of a ratio between the probability of the evidence 'given', or 'conditioned on', different scenarios. The likelihood ratio considers different, mutually exclusive hypotheses in different lines of the numerator and denominator.

The probability of the evidence is denoted as Pr (E). The prosecution hypothesis is denoted as 'Hp' and the defence hypothesis 'Hd', formulated as below.

$$\text{Likelihood ratio LR} = \Pr(E/Hp) \div \Pr(E/Hd)$$

Simply, the prosecution hypothesis Hp is that the person of interest contributed to the DNA profile, while the defence hypothesis Hd is that the person of interest did not – but an unknown person did. The likelihood ratio can become difficult to explain and interpret if:

- two or more contributors are considered;
- the evidence is conditioned on a known contributor such as a complainant, so that there are at least two contributors for both hypotheses;
- there are two suspects or more; and/or
- additional unknown/s.

These are deemed 'complex DNA profiles' and the propositions or hypotheses may be varied or multiple hypotheses need to be considered.

A likelihood ratio indicates if and to what extent the DNA analysis results support one proposition over another. It is not possible, on this basis alone, to determine which is the most probable proposition. To assign the probability of a proposition, the DNA analysis results should be combined with other information in the case. This is generally not considered to be the remit of the DNA scientist (Hicks et al., 2022).

An example of what the likelihood ratio is, and what it is not, is the Jama Case described in Chapter 1 (Case 1.3). The very high likelihood ratio for the DNA profile from the semen (source level) was seemingly transposed to the 'activity' of sexual intercourse with the complainant. The prior odds and posterior odds are a matter for the court but the likelihood ratio only referred to the DNA profile, a matter for the scientist. As it was found in the case, there *was* an error (contamination or inadvertent transfer in the collection). The semen did in fact come from the accused but the evidence had no meaning in the case context. The problem in the case was that the semen was the only evidence that a crime had occurred at all (offence level); it was 'sole plank' evidence.

There are no true likelihood ratios (Gill et al., 2018). Depending on assumptions, knowledge and the results to assess, different models will be adopted and hence different values for the likelihood ratio will be obtained. It is therefore important to outline in statements what factors impact evaluation – propositions, information, assumptions, data and choice of model (ibid.).

2.5.3 Transposition of the Conditional

The likelihood ratio statistic may increase error in thinking as the probability of the evidence (e.g., DNA profile) given the scenario may be transposed to state the statistic is the probability of the scenario given the DNA profile. The scientist is only concerned with the probability of the results – given the propositions – not with the probability of the propositions themselves.

A number of papers have emphasized this point.

> 'One cannot overstate the importance of distinguishing the probability of the DNA results given that the DNA came from an unknown person and the probability that the person is not the source of the DNA. When there is no other evidence in the case, if the court does not consider the other evidence (or the absence of it) in the case then even with very powerful results, there is a high risk of miscarriages of justice. Another danger is that the strength of the DNA results given source-level propositions is not equal to its value given activity-level propositions. The scientist should help explain the issue to the court by including a caveat in their statements'
>
> **(Gittelson et al., 2016 and quoted by Gill et al., 2018).**

It has been stated by statistics researchers in legal arenas that the fact that probabilistic fallacies continue to be advanced and considered in legal proceedings is an indictment of the lack of impact made by statisticians in general (and Bayesians in particular) on legal practitioners (Fenton et al., 2016).

Even forensic genetics papers that focus on DNA profiling state that avoiding the transposed conditional is not a straightforward endeavour and that many scientists are tempted to give an opinion on propositions. Not transposing the conditional is difficult and a lot of practice is needed. Respecting logic, forensic scientists should assign the probability of their findings given the truth of propositions, not the probability of the propositions given the findings (Hicks et al., 2022).

Courts of Appeal have transposed the conditional and these have been referred to in subsequent judgments so that there is a reinforcing of the fallacy. As an example, a ruling in a Supreme Court of Appeal in Victoria, Australia, stated:

> 'As explained by the (*trial*) court for each DNA sample where the suspect cannot be excluded as a contributor, a ratio was calculated which shows how much more likely it is that the suspect was the source of the DNA (or a contributor to it) than some other person chosen at random from the population was the source (or a contributor)'
>
> (*Tuite v. The Queen* [2015]; referred to in Hicks et al., 2022).

The above appeal ruling was referred to in another Supreme Court of Appeal in the same jurisdiction (*Davies v. The Queen* [2019]). Quoted at [187]: 'it was probative of the fact that a person 46,000 times more likely to be the applicant than a random member of the Australian Caucasian population had touched the bottle'. This is not only a transposition of the conditional but also an association error.

Another Appeal from the same jurisdiction had numerous association errors.

CASE 2.4 *VYATER V. THE QUEEN* [2020]

The DNA evidence challenged by the appellant concerned a glove found in a plastic tub, together with equipment which had been used for the manufacture of methylamphetamine. During the early morning of August 2016, police executed a search warrant on an upstairs office above a row of storage garage units. When police entered the premises the applicant was in the upstairs lounge area with a female acquaintance. The applicant gave evidence in his own defence. He acknowledged that he was the lessee of both premises but he did not live there although he stayed there on occasion, particularly on weekends when he would regularly consume methylamphetamine. The applicant told the jury that he had no involvement in the manufacture of methylamphetamine, although he was a user of the drug. As he lived at the premises, and was present when the police searched them, it was unsurprising to find his DNA transferred to multiple objects (this did not seem to be mentioned). Contamination mitigation procedures of the scene search were also not mentioned in the ruling (see Chapter 7 for importance of such).

The ruling stated: 'In the conventional manner, the evidence was presented in the form of a likelihood ratio. The likelihood ratio was unusually low. In the opinion of the witness, it was 26 times more likely that the DNA found on the glove came from the applicant than from a person selected at random'. This is stating the transposition of the conditional. The original trial had the scientific witness explain the DNA profile was a partial mixture originating from three contributors. 'The likelihood ratio showed ... that the DNA evidence is 26 times more likely if (the applicant) and two others are contributors than if three unknowns are contributors' as quoted in the Appeal ruling. Further, the original trial ruling was quoted 'means that roughly one person out of 26 has a DNA profile that matches the DNA on the left glove sample. This means, in a population like Australia of 25 million people, about a million people could have been responsible for that DNA profile that was picked up in the left glove'. Defence counsel put to the scientist that it was a '26 to 1' chance that it was the applicant's DNA on the red glove rather than that of an unknown person. The scientist agreed. This would mean, counsel then suggested, that if 100,000 people went through Flinders Street Station every day, then approximately 4,000 of them would give a similar DNA match. The scientist agreed that this was so, and that this meant 'in a sense' that a little under a million people in Australia would return a similar profile.

The Appeal ruling found there was no error in finding the DNA evidence admissible.

This author notes that one person cannot be responsible for a mixed DNA profile, the scientist transposed the conditional speculating on the scenario, and that likelihood ratios have been used in the jurisdiction since the 1990s (approaching 30 years), so it is disturbing that these fallacies perpetuate.

Note also that the staff DNA database likelihood ratio used to check for contamination in the examining laboratory has a threshold of 10,000. That is, likelihood ratios for staff members that are less than 10,000 are deemed to be adventitious, or match by chance. Consequently this text author queries the relevance of any likelihood ratios less than 10,000 produced in the jurisdiction.

2.6 Hierarchy of Propositions

2.6.1 Definitions

A DNA profile 'match', or a likelihood ratio of the DNA profile given certain alternative scenarios, may inform decisions about the DNA profile, but decisions about an activity – say sexual intercourse versus social contacts – involve additional considerations beyond the DNA profile. How and whether to statistically analyze more than just a DNA profile, especially when the forensic scientist may exist in a 'silo' situation remote from the crime scene analysis, is under debate. There is also the problem in conveying complex statistical analyses in the courtroom or to detectives prior to the prosecution stage. What is accepted is that the scientist should not relate a DNA profile, alone, to anything other than the statistical weighting of that profile.

The role of the forensic scientist is to provide an objective assessment of the evidence.

Interpretation of the evidence within a 'framework of propositions' describes the various 'levels' where the evidence may be evaluated (Gill, 2014). The framework provides a hierarchy where the value of the evidence increases at each level, according to a hierarchy of propositions. The 'evidence' may include all physical evidence located at a crime scene such as biological evidence. The hierarchy was first devised as an assessment tool in casework by the Forensic Science Service in England (Cook et al., 1998a and 1998b; Evett et al., 2002; Hicks et al., 2022).

Due attention must be paid to the position in the hierarchy of propositions that can be considered. This information must be effectively conveyed to the court to avoid the risk that an evaluation at one level is translated uncritically, and without modification, to evaluation at a higher level (Gittelson et al., 2016). The transposition of the conditional and association error may then occur (as discussed variously in this text).

The hierarchy of propositions is generally presented from source to offence and structured as a hierarchy as scientists need more information and knowledge to assess their results at activity level than at source level, for example. Experts should not stray outside their expertise when identifying the level where they can be most helpful (Hicks et al., 2022).

When the hierarchy of propositions was first devised, it was only possible to obtain a DNA profile from biological fluids present in relatively large amounts, such as a blood stain the size of a small coin. It has been described that in such cases it was reasonable to assume that the DNA profile was derived from a known biological fluid such as blood (Hicks et al., 2022).

The above presumption regarding source became questionable with the advent of more sensitive techniques, capable of analyzing smaller and smaller deposits to the point where invisible to the naked eye deposits were speculatively swabbed. This led to new levels of the sub-source level (the DNA profile level) and then the sub-sub-source level (the DNA profile contributor level) for situations where the issue is whether a person is part of the DNA mixture (e.g., a major contributor).

The hierarchy of propositions is described as follows, as a summary, in relation to DNA evidence (Gittelson et al., 2016; Gill et al., 2018, 2020; Hicks et al., 2022):

(a) Sub-source propositions consider the possibility that an individual is a source of the DNA from a trace, but do not infer body fluid or cell type. The results are the DNA profiles.

    With trace biological deposits that may be invisible and that cannot be attributed to a specific biological fluid or that has not been tested for such, the issue may only be: Who is the source of the DNA profile obtained? Sub-source level propositions may then be suitable. The resulting likelihood ratio obtained may be useful for providing investigative leads such as when there is no suspect and a DNA database search may be required.

(b) Source propositions consider the biological nature of the trace giving rise to the DNA profile and to whom it belongs. The trace is clearly defined as a particular body fluid, such as blood, saliva or semen, or as a particular cell type, such as epithelial cells. The results are the DNA profiles.

There should be no risk that the findings are misinterpreted with the alleged activities. This would typically be the case when the material is found in such a quantity that the material is relevant and the nature of the material can be safely assumed (such as a large amount of blood on a broken window).

If the nature of the material is disputed, it is more meaningful to consider activity-level propositions.

(c) Activity propositions consider the activities relating to how and when a trace was deposited and to whom it belongs. Typically the sub-source or source of the trace is not in dispute but only the mechanisms whereby the trace material was transferred.

Activity levels are used when the amount of collected trace material is low and considerations of transfer, persistence and recovery require specialized forensic knowledge. There is a widespread recognition that there is danger in leaving such assessments to non-forensic scientists and that it is the duty of the scientists to guide the court appropriately (Taylor et al., 2018a).

Sub-source level propositions might not be meaningful if the parties are known to each other or if the suspect has legitimate access to premises or objects, and DNA may be expected.

The evaluation involves extrinsic characteristics such as the quality of the DNA profile, relative quantity of DNA, where the DNA was sampled from, and should be considered when transfer, persistence or background have a significant impact on the understanding of the value of the findings in the context and relative activities.

'The real problems of interpreting poor quality traces and mixtures have only come to the fore in recent years. These problems have illuminated the important challenge that forensic science is facing: interpreting results in view of conflicting versions of events and activities' (Margot, 2011).

(d) Offense propositions are concerned with the issue of whether a person of interest is guilty or innocent of an offense.

The 'offense' level in the hierarchy of propositions (i.e., that the person is guilty of the offence or not guilty) cannot be addressed by the forensic scientist, it is a matter for the court.

See Figure 2.2 for a depiction of the hierarchy of propositions.

It is accepted that the higher up the hierarchy the propositions are, against which the scientists are competent to evaluate their results, the more directly useful the testimony will be to the court limiting the risk of misleading (Taylor et al., 2018a).

Because each case represents a unique set of circumstances, allegations (regarding posited activities) and results, it is difficult

FIGURE 2.2 The hierarchy of propositions.

to come up with a standard structure for evaluation. One common tool that assists in this task is Bayesian networks (BN), which are a graphical way of displaying and conducting complex probability evaluations (Taylor et al., 2018b). There is much diversity in the way that BN can be constructed.

2.6.2 Investigative and Evaluative Opinions and Reporting

Investigative versus evaluative opinions by forensic scientists in reports and testimony are considered to be an important distinction, in addition to the hierarchy of propositions. When there is no suspect, the forensic scientist works in an investigative manner. However, when there is a person of interest or defendant an evaluation by the forensic scientist is required to consider a framework of circumstances and case-specific propositions.

If there is new information, the scientist will need to re-evaluate the results. It is thus important that reports contain a caveat relating to this aspect (Gill et al., 2018).

2.7 Relevance of DNA to Criminal Investigation

DNA evidence may be obtained from visual biological fluids and/or identifiable body material such as blood or semen – but interpreted by authorities in a way seemingly ignoring the scientific method. The cases below consider transfer, explicitly or implicitly.

2.7.1 Ignoring Context

An interesting comparison can be drawn between Case 2.1 of David Camm and the tragic case of Kevin Brown, an analyst within a United States forensic crime laboratory who committed suicide due to his alleged association with a cold case murder. Both cases had two suspected offenders allegedly acting together with two different types of biological matter and two different DNA profiles.

CASE 2.5 KEVIN BROWN

A cold case investigation of the horrific murder of Claire Hough was started in 2012 in the San Diego Police Department crime laboratory. The 14-year-old girl had been found murdered, sexually assaulted and mutilated on a San Diego beach in 1984 (Davis, 2017; Flynn, 2020, Gotfredson, 2014; Larson, 2020). The crime laboratory found nothing of significance at the time and a vaginal swab was stored – the first investigation was conducted prior to the advent of DNA profiling.

The cold case investigation found blood staining on the girl's blue jeans with DNA that 'matched' the DNA of Ronald Tatro, a violent convicted rapist. The transfer of visible blood from Tatro to the girl's jeans was implicitly considered. Tatro had been on parole for a rape conviction at the time of the murder, a person of interest in the murder of a prostitute in San Diego and in 1985 convicted of the attempted rape of a 16-year-old girl. However, Tatro had drowned in a river in Tennessee in 2011.

There were also trace amounts of semen from the stored vaginal swab from the victim that contained DNA that matched Kevin Brown, a retired long-time employee of the laboratory. Police became fixated on the idea that Tatro and Brown were somehow involved together in the murder even though they had no known connections. Brown had sometimes been to 'strip' clubs, had depression and some people at the lab called him 'kinky'. During January 2014, police confiscated many items from the home of Brown (14 boxes) amid relentless questioning and with no result. Ten months later Kevin Brown was found hanged in a tree by a road near a state forest. Four days after his death police identified him publicly as a suspect in the Hough murder.

It was common practice for laboratory employees in the 1980s in San Diego to provide intimate samples such as semen to be used as test controls, such as a positive control for the screening test for semen seminal acid phosphatase (from this author's experience I can confirm this happened in Australia as well). Controls and other samples were often left exposed and vulnerable to contamination due to a lack of awareness of the ready transfer of DNA material. Also, gloves were not routinely changed. Vaginal swabs would be screened for semen and then stored.

During 2015, Kevin Brown's widow sued the city of San Diego and police for the wrongful death of her husband. A jury in 2020 awarded Rebecca Brown more than 6 million dollars for illegal search and seizure and abuse of power, and ordered the investigating detective to personally pay 50 thousand dollars in punitive damages.

Staff DNA databases are used to eliminate staff from any potential contamination events, as it is recognized that their DNA may be present (through no matter what way) at any place in the laboratory. Staff databases are not used as a potential pool of suspects for a crime. Staff databases are later discussed in Chapter 7. Further, 'cold cases' were examined originally in conditions that pre-dated the high sensitivity era of current DNA profiling and unaware of DNA contamination mitigation procedures. The notion that laboratory contamination is not possible has been disproved due to numerous instances of known contamination.

2.7.2 Undermining DNA Evidence

There have been cases where DNA evidence in a case has been undermined by prosecuting authorities in an effort to support their original suspect. One hypothesis proposed by the prosecution in one of the numerous proceedings against the accused in the below case contradicted the rationale traditionally used in sexual offence cases. This is a type of cognitive bias or cognitive dissonance whereby the evidence contradicts the original theory but is nonetheless ignored or dismissed. The case follows (Innocence Project; Martin, 2011).

CASE 2.6 JUAN RIVERA

During 1992 in the United States, an 11-year-old girl was raped and murdered. Holly Staker was babysitting two younger children – 2 and 5 years old, in Illinois. Juan Rivera was convicted for the murder in 1993. He was 19 years old at the time and had previously been convicted of burglary and was on home monitoring; he broke down and confessed after days of relentless questioning. He had a further two trials, each after an appeal, and was convicted each time. During 2005 there was DNA testing of the semen found inside the vagina of the deceased which excluded Juan Rivera and a 2006 court vacated the conviction and ordered a new trial. A jury trial in 2009 subsequently convicted Rivera and he was sentenced to life in prison. An Appellate Court in 2011 ruled that the conviction could not stand. Juan Rivera was released from prison in 2012.

During the third trial in 2009 the police and prosecutors had explained to the court that the DNA obtained from the semen from Holly Staker was not relevant to Rivera by presenting two theories. One was contamination via some means, and the other theory was previous sex with another man (both theories considering transfer as an implicit issue). Regarding the contamination theory, where the prosecution speculated that no sperm was recovered from the autopsy, but someone else's sperm came in contact with the swab sometime later, the later Appellate Court dismissed it as 'highly improbable'. The State did not present any evidence that the victim was in a relationship with anyone. The most reasonable explanation, therefore, of who murdered the victim was not the defendant but rather someone who, unfortunately, had not yet been identified.

A federal lawsuit against wrongful conviction was settled in March 2015 where Rivera was awarded $20 million (National Registry of Exonerations, 2022).

Considering the propositions at the activity level would not have resolved this situation, since the prosecution seemed to consider the DNA results irrelevant or otherwise explained away.

2.8 Relevance of DNA Transfer to Criminal Investigation

The first cases involving DNA analysis concerned what was considered a 'direct DNA transfer' of the questioned visible and identifiable material, such as in Case 2.2 of this chapter. 'Indirect transfer' was traditionally considered as contamination of that material exhibit in question when the context of the case was considered.

As the DNA analysis techniques became more sensitive, and the results of smaller and smaller samples resulted in interpretable DNA profiles, the desire to obtain information from deposits that could not be related to a biological material such as blood or semen grew.

A DNA profile may also be conflated with the action of transfer from an individual. The term 'touch DNA' may have contributed to the idea of a DNA profile obtained via direct transfer. There are many steps involved in the process of a person transferring DNA existing in their body to an item and any subsequent production of a DNA profile – each step requires considerable attention to detail in analysis and interpretation.

The first step is *transfer* from the body – the DNA material itself – and can be directly or indirectly to a surface. Second is the *persistence* of that deposit, which may be subject to environmental conditions or other impacts. Third, *prevalence* of DNA on and in the environment of the exhibit, which may include multiple DNA deposits at various times and from various people that may add complexity to the interpretation. Finally, *recovery* of the deposit, which includes searching/locating, sampling, extraction and analysis.

Visible deposits of biological material such as blood staining have different considerations to non-visible trace DNA transfer. One example is a full fingerprint in blood where a complete DNA profile associated with the blood can be determined and the fingerprint has a full enhancement. The fingerprint is complete and thus readily searchable to a database; similarly the DNA profile can be compared to a reference sample or DNA database with confidence. Contrast this to a non-visible deposit from an item, obtained through speculative swabbing, and a partial, mixture DNA profile from the swab. The limitations in the latter evidence should be fully explained in any forensic report detailing results.

Much more needs to be done to resolve the current paucity of empirical data on the variables that may impact DNA trace transfer (van Oorschot et al., 2019). Non-visible deposits of trace DNA material have extra considerations to visible deposits of biological fluid such as blood. Hands, for example, can be a major and highly contested source of DNA-containing material detected at crime scenes (ibid.).

DNA trace transfer has become a discipline of its own and 'DNA-TPPR' involves the understanding of transfer (T), persistence (P), prevalence (P) and recovery (R) issues. As described in

Chapter 1, this type of terminology was derived from investigating visible trace transfer of matter. It is important to understand the difference between visible and non-visible material and the layer of uncertainty that is added with investigating non-visible material.

The limitations of much of the DNA transfer related research studies conducted to date have been highlighted in a review paper (Gosch and Courts, 2019). Of particular relevance to case studies examined in this book are the multitude of factors that need to be considered and that these may be underestimated or oversimplified in research studies. Consequently extrapolation from transfer experiments in a laboratory need careful thought before the reporting scientist applies such findings or principles to a criminal case at hand.

The following chapters will discuss the above varied issues.

References

Aitken, C., Roberts, P., and Jackson, G., 2010, Practitioner guide 1 – Fundamentals of probability and statistical evidence in criminal proceedings, Communicating and interpreting statistical evidence in the administration of criminal justice, Royal Statistical Society, London, available at www.rss.org.uk/statsandlaw

Baeschler, S., 2016, Study of criteria influencing the success rate of DNA swabs in operational conditions: A contribution to an evidence-based approach to a crime scene investigation and triage, *Forensic Science International: Genetics*, 20, 130–139.

Biedermann, A., Champod, C., Jackson, G., et al., 2016, Evaluation of forensic DNA traces when propositions of interest relate to activities: Analysis and discussion of recurrent concerns, *Frontiers in Genetics*, 7, 215.

Cook, R., Evett, I., Jackson, G., et al., 1998a, A hierarchy of propositions: Deciding which level to address in casework, *Science and Justice*, 38, 231–39.

Cook, R., Evett, I., Jackson, G., et al., 1998b, A model for case assessment and interpretation, *Science and Justice*, 38, 151–56.

Davies v The Queen [2019] Supreme Court of Appeal, Victoria, VSCA 66.

Davis, K., 2017, Judge won't toss lawsuit accusing San Diego police of driving ex-criminalist to suicide, *The San Diego-Union Tribune*, May 26.

Evett, I., Gill, P., Jackson, G., et al., 2002, Interpreting small quantities of DNA: The hierarchy of propositions and the use of Bayesian networks, *Journal of Forensic Sciences*, 47, 520–530.

Fenton, N., Neil, M., and Berger, D., 2016, Bayes and the Law, *Annual Review of Statistics and Its Applications*, 3, 51–77.

Flynn, M., 2020, Police suspected a crime lab technician of murder. Their mistake led him to hang himself, his widow says, *The Washington Post*, February 5.

Gill, P., 2014, *Misleading DNA evidence: Reasons for miscarriage of justice*, Academic Press Elsevier, London and New York.

Gill, P. and Werrett, D., 1987, Exclusion of a man charged with murder by DNA fingerprinting, *Forensic Science International*, 35, 145–148.

Gill, P., Brenner, C., Buckleton, J., et al., 2006, DNA commission of the international society of forensic genetics: Recommendations on the interpretation of mixtures, *Forensic Science International*, 160, 90–101.

Gill, P., Gusmao, L., Haned. H., et al., 2012, DNA commission of the international society of forensic genetics: Recommendations on the evaluation of STR typing results that may include drop out and/or drop-in using probabilistic methods, *Forensic Science International: Genetics*, 6, 679–688.

Gill, P., Hicks, T., Butler, J., et al., 2018, DNA commission of the international society for forensic genetics: Assessing the value of forensic biological evidence – Guidelines highlighting the importance of propositions, Part I: evaluation of DNA profiling comparisons given (sub)-source propositions, *Forensic Science International: Genetics*, 36, 189–202.

Gill, P., Hicks, T., Butler, J., et al., 2020, DNA commission of the international society for forensic genetics: Assessing the value of forensic biological evidence – Guidelines highlighting the importance of propositions. Part II: Evaluation of biological traces considering activity level propositions, *Forensic Science International: Genetics*, 44, 1–13.

Gittelson, S., Kalafut, T., Myers, S., et al., 2016, A practical guide for the formulation of propositions in the Bayesian approach to DNA evidence interpretation in an adversarial environment, *Journal of Forensic Sciences*, 61, 1, 187–195.

Gosch, A., and Courts, C., On DNA transfer: The lack of systematic research and how to do it better, *Forensic Science International: Genetics*, 40, 24–36.

Gotfredson, D., 2014, CBS News 8 investigates: The death of Ronald Tatro, *CBS News online*, November 21.

Gupta, S., 2017, Written in blood, *Nature*, 549, September, 524–525.

Hicks, T., Buckleton, J., Castella, V., et al., 2022, A logical framework for forensic DNA interpretation, *Genes*, 13, 957.

Human Genome Project, available at http://www.ornl.gov/sci/techresources

Inman, K. and Rudin, N., 2001, *Principles and practice of criminalistics: The profession of forensic science*, CRC Press, Boca Raton, Florida.

Innocence Project, available at http://www.innocenceproject.org

Larson, T., 2020, *Tales of true crime, episode 17: DNA gone wrong and the suicide of Kevin Brown*, KFGO, North Dakota, March 9.

Margot, P., 2011, Commentary on: The need for a research culture in the forensic sciences, *UCLA Law Review*, 58, 795–801.

Martin, A., 2011, The prosecution case against DNA, *New York Times Magazine*, November 27.

Meakin, G., Kokshoorn, B., van Oorschot, R., et al., 2021, Evaluating forensic DNA evidence: Connecting the dots, *WIREs Forensic Science*, 3, 4, July/August, e1404.

Morris, E., 2012, *Inquest into the death of Azaria Chamberlain* NTMC 020.

Morling, T.R., 1987, *Royal commission of inquiry into Chamberlain convictions*, Government Printer of the Northern Territory, Darwin.

National Academy of Sciences, 2009, *Strengthening forensic science in the United States: A path forward*, National Academies Press, Washington, DC.

National Registry of Exonerations, 2022, available at http://www.law.umich.edu

Nurk, S., Koren, S., Rhie, A., et al., 2022, The complete sequence of a human genome, *Science*, 376, 44–53.

Peel, C. and Gill, P., 2004, Attribution of DNA profiles to body fluid stains, *International Congress Series*, 1261, 53–55.

Reynolds, M. and Silieniks, T., 2016, Considerations for the assessment of bloodstains on fabrics, *Journal of Bloodstain Pattern Analysts*, 32(2), 15–20.

Sari, D., Hitchcock, C., Collins, S., et al., 2020, Amylase testing on intimate samples from pre-pubescent, post-pubescent and post-menopausal females: Implications for forensic casework in sexual assault allegations, *Australian Journal of Forensic Sciences*, 52, 6, 618–625.

Schneider, G., 2013, *Expert testifies in David Camm murder trial that 'determining blood stains' cause has 50% error rate*, October 4, Courier-jounal.com.

Taupin, J.M. and Cwiklik, C. 2010, *Scientific protocols for forensic examination of clothing*, CRC Press, Boca Raton, Florida.

Taylor, D., Abamo, D., Rowe, E., et al., 2016, Observations of DNA transfer within an operational forensic biology laboratory, *Forensic Science International: Genetics*, 23, 33–49.

Taylor, D., Kokshoorn, B., and Biedermann, A., 2018a, Evaluation of forensic genetics findings given activity level propositions: A review, *Forensic Science International: Genetics*, 36, 34–49.

Taylor, D., Biedermann, A., Hicks, T., et al., 2018b, A template for constructing Bayesian networks in forensic biology cases when considering activity level propositions, *Forensic Science International: Genetics*, 33, 136–146.

Taylor, M., 2022, Study: Forensic STRs may reveal medical information, *Forensic News*, October 3.

Tuite v The Queen [2015] Supreme Court of Appeal, Victoria, VSCA 148.
van Oorschot, R., Szkuta, B., Ballantyne, K., et al., 2017, Need for dedicated training, competency assessment, authorizations and ongoing proficiency testing for those addressing DNA transfer issues, *Forensic Science International: Genetics Supplement Series*, 6, e32–e34.
van Oorschot, R., Szkuta, B., Meakin, G., et al., 2019, DNA transfer in forensic science: A review, *Forensic Science International: Genetics*, 38, 140–166.
Vyater v The Queen [2020] Supreme Court of Appeal, Victoria, VSCA 32.
Wambaugh, J., 1989, *The blooding*, William Morrow, New York.

Chapter 3

TRACE DNA

BOX 3

- Definition of trace and 'touch' DNA.
- Transfer of trace DNA.
- Speculative detection and recovery of trace deposits.
- Sampling rationale and methods for trace DNA.
- Relocation of trace DNA.
- Context of trace DNA.

3.0 Introduction

It has been stated that nowadays the attribution of a DNA trace to an individual is only rarely questioned. This is due to the power of DNA-based individualization where the presence of DNA-containing material from a specific individual may be proven beyond rational doubt (e.g., Gosch and Courts, 2019; Helmus et al., 2020), although this has been clarified to mean a single-source DNA profile of adequate quality (Otten et al., 2019). When

DOI: 10.4324/9781003158844-3

considering mixed DNA profiles and low-quantity/-quality DNA profiles, this may not be the case. DNA is a powerful discrimination tool, but all scientific methods have their limits and no scientific technique is error free.

If the donor/s of the DNA deposit is accepted by all parties in the judicial system, then the debate may swing to, 'How did the DNA get there?' The discriminating power of DNA profiling does not allow for contextualization of the trace material and thus to determine how a particular deposit of DNA got to the area from which it was sampled.

Expert witnesses in court are often asked to address questions that concern the activity which led to the deposition of a defendant's DNA on an item or surface (Johannessen et al., 2021). In particular, if the source of the DNA is considered to be from biological material deposited after skin contact, it is common that the defendant has an alternative explanation as to how their DNA was deposited which often involves indirect transfer of DNA (ibid.). This text's author questions whether this is considered to apply to the social or domestic associates of the complainant (which is a reasonable alternative explanation as this book demonstrates) or to 'unknown' individuals to the complainant. It is also important to clarify whether visible identified biological fluids, or trace DNA, is the material considered.

It has also been stated that the most challenging cases regarding *probabilistic* approaches to evaluating evidence – given 'activity' level propositions – are those involving trace DNA (Gill et al., 2020). This is because of the ready propensity for trace DNA to be transferred and the many factors to factor concomitantly.

Further, it is not possible to determine whether the trace DNA has been transferred through direct means or indirect ways. This author has reviewed cases in recent years where a list of variables affecting trace DNA transfer – such as surface on which the material is found (smooth or porous, for example) – is provided in the body of the laboratory report, the appendix or in testimony. Such commentary delves into 'activity' level propositions without necessarily being explicit that this is the case. There have also been cases where the testifying scientist has stated, 'in my opinion the trace DNA was deposited through direct transfer rather than secondary transfer', thus providing an explanation for the results, not an evaluation of the results in the light of the different scenarios (see Chapter 2).

A DNA profile may be conflated with a direct deposit of DNA from an individual, or the action of transfer from an individual. There are many steps involved in the process of a person transferring DNA from their body to an item and the subsequent production of a DNA profile – each step requires considerable attention to detail.

The first step is transfer from the human body – the DNA material itself – and can be direct or indirect to the item. Second is the persistence of that deposit, which may be subject to environmental conditions or other impacts such as cleaning. Third, prevalence of DNA on and in the environment of the exhibit, which may include multiple DNA deposits at various times and from various people that may add complexity to the interpretation. Finally, recovery of the deposit, which includes searching/locating, sampling, extraction and analysis. (See Figure 3.1.)

Research on DNA transfer has shown that the transfer, persistence and recovery of *trace* DNA is a highly complex and multifactorial process. Accordingly a multitude of parameters need to be considered when evaluating how a trace DNA sample result

Variables in the production of a DNA profile from a person.

1. Transfer from the body; related to the source such as blood, sperm, saliva, body secretions, epithelial cells and extracellular DNA

Direct or indirect transfer via intermediary items

2. The exhibit in question; may be transferred as a stain, surface material, co-mingled with other matter
3. Persistence of DNA material
4. Prevalence of DNA, including other self and non-self DNA
5. Locating the DNA (recovery techniques)
6. Sampling
7. Extraction of DNA from material sampled
8. Quantification of DNA
9. Amplification
10. Production of a visual diagrammatic representation of a DNA profile

FIGURE 3.1 Obtaining a DNA profile.

may have occurred from a suggested scenario or otherwise (e.g., Gosch and Courts, 2019).

3.1 Definition of 'Trace DNA' versus 'Touch DNA'

Trace DNA is defined in this text as DNA that cannot be related to a specific biological origin or matter, such as from blood, semen, saliva or skin cells. It has replaced the term 'touch DNA' as it cannot be assumed that trace DNA amounts originated from 'touching' with the skin of the hand or other skin contact (e.g., Atkinson et al., 2022; Gosch and Courts, 2019; Meakin and Jamieson, 2013).

DNA deposited as a trace sample may have been directly or indirectly deposited at the crime scene. The recovery of directly deposited trace DNA varies depending on both the substrate (material on which the DNA was deposited) and exposure or contact time, with absorbant substrates yielding a greater quantity of DNA (Meakin and Jamieson, 2013; Gosch and Courts, 2019). Indirect deposition has been shown to yield similar quantities of DNA as direct deposition.

Trace DNA is not visible and/or testing for body origin has not been performed or been unsuccessful. There is also currently no test used for skin cellular presence at the crime scene, or within a medical examination room or forensic laboratory. A 'swab it and see' or speculative approach is used to determine if DNA is present in a non-visible deposit. It is not possible to determine the quantity or quality of any DNA that may be present before the actual analysis.

Due to the increasing reliance on DNA as a forensic tool, there has been a desire to generate information from ever smaller amounts of DNA. Sometimes a DNA profile can be generated when no DNA is quantified. Trace DNA samples had previously been described as any sample which falls below recommended thresholds at any stage of the analysis, from sample detection through to profile interpretation, and cannot be defined by a precise picogram amount (van Oorschot et al., 2010).

It was first reported in 1997 that DNA profiles could be obtained from touched objects (van Oorschot and Jones, 1997). This discovery was initially met with disbelief by many within the forensic community (van Oorschot et al., 2019). However, once verified, the finding that enough measurable but non-visible DNA material could be deposited by holding an item with a hand, and the extrapolation of this observation to contact with skin in general, drastically broadened the types of cases in which DNA profiling could be applied (ibid.). The discovery led to the collection of DNA from a wider range of exhibits such as tools and weapons, clothing, windows and even stones – as well as expanding the types of offences traditionally examined with forensic DNA (i.e., serious crime) to volume crime.

The early years of trace DNA profiling had generated debate regarding the difficulties that a significant increase in sample submission brought to forensic laboratories. Later, laboratories used the favourable prospect of trace DNA to argue for increased staff and budgets and improved laboratory facilities (van Oorschot et al., 2010). Samples collected from potentially 'touched' objects now represent more than half the total number of samples processed for DNA profiling (van Oorschot et al., 2019).

The increased sensitivity of DNA profiling kits and the types of objects from which samples are collected have meant that many DNA profiles are partial – fragments are missing due to insufficient quality samples for amplification needs. Many DNA profiles are also a mixture of multiple contributing individuals, sometimes four or more, together in the one profile. Potential scenarios as to how a particular DNA contributor led to the transfer and detection in the sample have also increased together with the knowledge of how DNA may transfer.

'Touch' DNA is a term commonly used in casework and in the literature for the examination of objects potentially having deposits of DNA through handling. This generally refers to trace DNA where the DNA has not been identified as to body origin, or low amounts of DNA. Describing such a sample as 'touch DNA' however can be misleading (van Oorschot et al., 2019; Gill et al., 2020). A specific mode of action is implied and also a biological source (from skin). Even if the source and/or actions are known – such as in research experiments or mock simulations – it cannot be assumed that the cells transferred came from the hand itself or even skin on the body from the person in question unless the experiments are strictly controlled. The hand or skin itself may have pre-existing deposits of biological material from other regions of the body, or DNA material from other people.

As examples, this author has reviewed cases where the terms 'contact DNA', 'skin DNA' and 'touch DNA' have been used by either police or forensic personnel for invisible deposits speculatively swabbed for DNA. Such terms imply actions or body source when, in fact, these factors may be unknown even after forensic examination. The courts may debate the sample denoted as such without being aware that the sample may consist of other DNA material than skin cells and may be deposited in other ways than 'contact'.

The connection of trace DNA with 'touched' objects since the paper published in *Nature* in 1997 may have led to the term 'touch DNA' instead of the proper terminology of 'trace' DNA. There have been innumerable articles published since then describing DNA transfer via surface to surface with no contact via the hand or body. There have also been recent data describing DNA transfer from object to object with no contact, such as shaking an item over a surface. The term 'touch DNA' is not advocated by this author and when used should be viewed with additional concern as to the meaning.

It has been previously noted (Meakin and Jamieson, 2013; Gill, 2014) that when considering trace DNA and its transfer, the general limitations of trace DNA should apply. These limitations are:

- the type of cell from which it derived and when it was deposited is unknown;
- it is not possible to make conclusions regarding transfer and persistence; and
- because the test is sensitive it is common to encounter mixtures, but it is unknown if any DNA is relevant to the case.

A 'statement of limitations' should apply when 'activity' level propositions have not been addressed in the case and there is 'trace DNA' evidence (Gill, 2014). Miscarriages of justice occur when it is implied that the evidence has more meaning (i.e., certain assumptions) than it really does. All the different possibilities describing how a 'trace DNA profile' may be transferred should be listed (ibid.):

1. The trace DNA profile was transferred during the crime event.
2. The trace DNA profile was part of the background contamination of the crime scene (innocent transfer), and
3. The trace DNA profile was a post-crime scene contamination event (investigator-mediated contamination).

It was noted in 2010 that working effectively with trace DNA samples means considering all aspects of the process rather than simply focusing on the interpretation phase (van Oorschot et al., 2010). An understanding of factors relating to collection, extraction, amplification and interpretation as well as issues relating to contamination and transfer is required. This aspect is no different to any proper sampling and testing plan and rationale using the scientific method as required in the examination of exhibits (Taupin and Cwiklik, 2010).

Complicating factors, when examining 'trace' DNA samples that may not be visible, are the influences of the recovery and analysis procedures (and indeed collection and handling). It has been observed that the influence of the recovery and analysis procedures of trace DNA, including location of the sample, is faintly reminiscent of Heisenberg's uncertainty principle – one of the most famous principles in physics which states that there is a 'fuzziness' in nature and a fundamental limit in what we can know about the smallest scales of nature. Regarding forensic science detection, the object of interest can never be observed exactly as it is changed by the very procedure of observation (Gosch and Courts, 2019). The large inter-laboratory variability regarding protocols for the recovery and analysis of biological material from an exhibit demonstrates the problem.

This aspect also influences trace DNA transfer research as the variance caused by variability in collection practices, DNA extraction methods and so on may be larger than variables studied to influence transfer, thus effectively impeding comparison between studies and assessment of the true transfer variance in probabilities (ibid.).

When assessing the value of trace DNA, there are attempts in the forensic community to determine 'activity level' propositions (probability of observing the evidence given different activities) incorporating DNA transfer and its associated concepts. Believed by many as the most appropriate approach to dealing with small quantities of DNA (Butler and Willis, 2020), effectively changing the DNA results from the utility of its discrimination power of DNA profiling.

This field of endeavour will be discussed in the following chapters. It is a complex area, especially when relating experimental data or literature data to a specific 'event' of a crime. The reporting scientist is effectively performing a 'crime scene reconstruction', when possibly that scientist has never been to the crime scene or even examined the exhibit/s or photographs of them (see Chapter 1 for the limited information that forensic DNA analysts may receive).

3.2 Trace DNA Transfer

3.2.1 History

The principles of transfer of *trace* DNA are similar to that of any forensic *trace material* such as fibres, paint and glass based on Locard's Theorem, or Locard's Exchange Principle, as the principle of trace evidence transfer, described in Chapter 1. Whether that trace is *detectable* depends on the quantity and quality of the material and analytical methods used in detection and sampling. The *persistence* of the trace on items depends on the activities involving the trace, and the environment in which the activities occurred, after the trace was deposited.

Due to the often-non-visible deposits and/or limited quantities with trace DNA, there are additional complexities involved at all stages of the analysis, starting from collection at the crime scene to the final communication of the results to the criminal justice system.

A trace DNA deposit may arrive from *direct* transfer at a time previous to the questioned event as well as during the event, or from multiple direct transfer events. It may also arrive through *indirect* transfer, via multiple ways and multiple events. It is also not possible to determine from a DNA profile, or the quantity of DNA obtained, whether the deposit consists of a combination of direct and indirect transfer events.

A biological substance that has been transferred multiple times, if detectable, may appear as components of complex DNA profiles. This is because the *vectors* (such as hands or implements) aiding the transfer, and/or the *substrate* from which it is ultimately collected, may also contain DNA (Goray et al., 2010).

The first study on trace DNA transfer was in Melbourne, Australia. It showed that DNA can be recovered from objects that had been touched by hands (van Oorschot and Jones, 1997). The finding regarding 'touched' objects was heavily publicized in the media as well as the forensic community due to its investigative potential (author experience working in the particular laboratory at the time).

The study had volunteers handling plastic tubes; the tubes were then swabbed, and DNA profiles were observed that matched the holder of the tube. This was in 1997 when less-sensitive DNA testing kits were being used compared to current kits.

The other findings in this study are just as valuable. Volunteers handling tubes had their hands swabbed and DNA profiles were observed that matched the previous holders of the tube – the volunteers had not contacted each other. Tubes held for a short time by a second or third person usually provided the DNA profile of the last holder but also provided the DNA of previous holders. The study showed that:

- DNA may be transferred from hand to object (direct transfer) and then from object to hand (secondary transfer).
- There may be no physical contact between the original depositor and the final surface on which the DNA profile was located (indirect transfer).
- DNA yields from tubes held for varying lengths of time (5 seconds, 30 seconds, 3 minutes and 10 minutes) did not vary significantly indicating substantial transfer during initial contact.
- Hands swabbed before and after a 1-minute-long handshake revealed the transfer of DNA from one individual to another in one of the four hands tested – thus DNA could be transferred but was not always transferred (or at least detected).

- Genetic profiles from objects handled by several people or from minute blood stains on touched objects may be difficult to interpret.
- There is a need for caution when handling exhibits and interpreting results.

The preceding bullet points and their relevance to case work took longer to appreciate in the forensic community. It has only recently been recognized that more needs to be done to understand the variables associated with the indirect transfer of DNA (van Oorschot et al., 2019).

3.2.2 Terminology of DNA Transfer

Direct and indirect transfer relates to the ways or routes by which DNA may be transferred. The terms 'direct transfer' and 'primary transfer' mean that an individual directly deposits their DNA onto a surface (this includes without contact such as spitting or coughing). 'Indirect transfer' means that there has been an intermediary surface/s between the initial deposit and the final collection of a sample.

3.2.2.1 Direct Transfer

An older review that is still applicable (Meakin and Jamieson, 2013) has described that it is possible to touch an item once and leave no detectable DNA, or leave a relatively large amount of DNA. Regular use of an item may also leave variable quantities of DNA. The review states that it is thus not possible to establish from the amount of DNA recovered from a surface whether the DNA was deposited by a singular touch or by regular use. It also states that it is not possible to use the quality of a DNA profile to establish whether the DNA came from the last handler.

Direct transfer of trace DNA does not mean that deposit is relevant to the case at hand. Examples of direct transfer of DNA, misinterpreted as having meaning, are frequently found in the literature on the subject and can be considered as 'contamination' – DNA that has been transferred subsequent to a crime event.

A high-profile example, which led to an international ISO standard for DNA collection tools, is described below.

CASE 3.1 THE PHANTOM OF HEILBRONN

The 'Phantom of Heilbronn' – also known as 'the woman without a face' – was believed to be Germany's most dangerous woman at the time during the 2000s.

The 'Phantom' was believed to be not only responsible for at least six murders but also a common thief. Forty unsolved crimes over 15 years (from 1993) in Germany and Austria were linked by DNA only. These included a high-profile 2007 murder of a young policewoman Michele Kiesewetter in Heilbronn, a town in southern Germany, the locale of the crime giving the unknown offender the name. The Phantom's female DNA was found at a car dealership burglary as well as a school break and enter, but in both cases her convicted 'accomplices' denied her existence. Her DNA (from non-visible deposits of trace DNA) was found on items as various as a tea cup, a cookie, a heroin syringe, on a fired bullet in a feud between two brothers and on the surfaces inside a car and on a petrol cap on another car. No security camera had ever captured her image (Temko, 2008).

More than 100 police and prosecutors in Germany, and backed by officers in Austria and France, were involved in the investigation, spending 8 years and an estimated €2 million. The high level of publicity and devotion of police resources was socially alarming. Saliva swabs for reference DNA were taken from nearly 3000 homeless women, or those women deemed to be of dubious character, from places as various as southern Germany, France, even Belgium and Italy – but there was no 'match'.

Some of the cases were 'cold cases' that had been re-investigated using more sensitive DNA technology. Trace DNA was collected from an exhibit with 'swabs' typically composed of small wooden sticks with a wad of cotton wool at the end (similar to cosmetic 'cotton buds'). The swab ends were moistened with water and then wiped along the suspected surface that may have the deposit of DNA to collect any material – thus a speculative search.

The real 'Phantom' was discovered when officials were trying to establish the identity of a burned corpse at the German/French border – neither fingerprints nor DNA could be obtained from the body. A swab taken from an associated application form from a male asylum seeker, who disappeared in 2002, was found to contain the female Phantom's DNA (Diehl and Juttner, 2009). This was thought impossible as the form was supposedly from a male

> person, so examiners repeated the analysis with another cotton swab – to find that the DNA was not there. It was ultimately discovered that the cotton swabs used to collect the relevant crime scene samples were contaminated. The origin of the DNA was eventually traced back to an innocent woman working in a cotton swab packaging factory in Bavaria (Spiegel online, 2009). It led the German newspaper 'Bild' to headline 'Are the heads of our police stuffed with cotton wool?' (Himmelreich, 2009).

Police had for years believed that the traces of DNA taken from very different crime scenes belonged to a single female offender. However, investigators had brought the DNA to the crime scene, or at least the exhibits taken from them, by using contaminated swabs. The concern this prompted led to an international scientific standard in 2016 to describe DNA collection tools as 'DNA free' rather than just 'sterile' as denoted by the manufacturer (ISO Standard 18385:2016, available at www.iso.org).

The murder of the policewoman was eventually attributed to a neo-Nazi terrorist cell believed to have also committed another nine murders known as the 'Bosphorus murders' (Spiegel online, 2012). Police had assumed that these murders were committed by the Turkish mafia fighting turf wars – giving the name 'Bosphorus' – but instead were later connected to the 'National Socialist Underground' (Meaney and Schafer, 2016).

The 'Phantom' highlights that the use of sensitive DNA technology has accompanying assumptions and limitations, and requires recognition and mitigation of potential contamination at each step of the process between collection, analysis and interpretation. Thus, sometimes the question is not just 'Whose DNA is this?' but also 'How did the DNA get here?'

3.2.2.2 Indirect Transfer

Indirect transfer incorporates all the multiple transfer events that may occur after an initial deposit. 'Secondary transfer' is the second step after the first or primary transfer. Tertiary transfer is the third following step after secondary transfer. Quaternary transfer is the next step, and there may be further steps such as quinary (five times), senary (six times) and so on. (See Figure 3.2.)

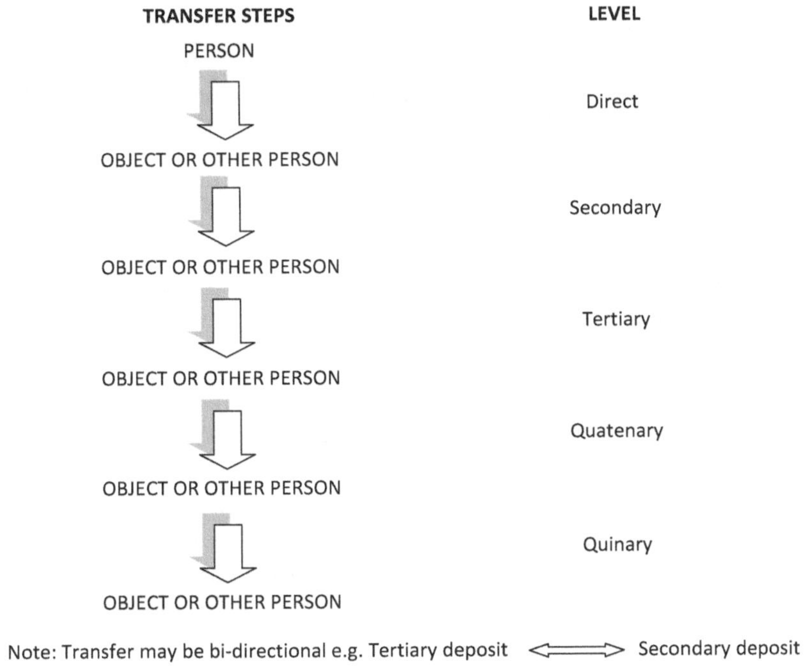

FIGURE 3.2 Transfer steps.

Indirect transfer is frequently referred to in the literature and in case studies, court judgments and commissions of inquiry. There have also been notable instances in the literature, and also case studies, that describe 'tertiary transfer' and 'quaternary transfer', from the early days of knowledge of DNA transfer to current literature.

Levels of DNA transfer higher than tertiary (derived through at least two intermediary items) were postulated in an early study from 2006, using less-sensitive technology than today (Poy and van Oorschot, 2006a). The study was in the Australia, Victoria laboratory that first documented the concept of DNA from handling in 1997. The extent and origin of DNA material was investigated and swabs or tape lifts from items on or around an examination bench were taken in an attempt to determine *background* levels of DNA in the laboratory.

A magnification lamp in the laboratory yielded an almost full DNA profile and matched a number of samples on a DNA database, related to the one case. It was discovered that one blood stain and one trace sample were taken originally from the same item, described as a 'bulky jacket', 3 months prior to the swabbing of the magnifying lamp. The study suggested that the DNA appearing on the top side of the lamp was due to transfer from the jacket onto a glove (secondary transfer) and then from the glove onto the lamp (tertiary transfer) when the jacket was illuminated with the lamp. A quaternary and even higher transfer of DNA may have occurred subsequently. This was through transfer from jacket to glove (secondary) to lamp (tertiary) to other glove (quaternary) to another exhibit (fifth substrate).

Another finding of these same authors (Poy and van Oorschot, 2006b) concerned surfaces in the laboratory considered to pose a low risk for contamination, including floors and taps. These were thought to be low risk because multiple transfer steps from an examination would be required to produce a DNA profile from that item. However, five out of six 'low risk' items provided a DNA profile, but the origin of these profiles remained unknown. This finding demonstrates the prevalence of DNA in the laboratory, even on items deemed to be low risk.

A 2015 Australian study (Goray and van Oorschot, 2015) examined unscripted social interactions between three individuals having a drink at a table for the duration of 20 minutes. The analysis showed that simple everyday interactions involving only a few minor items in some instances led to detectable DNA being transferred among individuals and objects without them physically having contacted each other, through secondary and higher transfer. There were several instances where the DNA transferred was the major component of the profile. In some instances the participants acted as vectors for foreign DNA; non-self DNA that may have already been present on the hands. This also included 'foreign' DNA from unknown individuals; in 30% of samples a DNA profile was obtained which excluded all participants. A jug handle had a mixture of DNA of 6 individuals, and one hand had an un-interpretable mixture of DNA from a proposed 12 individuals (there were 3 individuals in the experiment).

This study showed that humans are very tactile and will often touch themselves and objects over a short period of time.

When individuals have been in direct or indirect contact with each other, such as in social or domestic situations, the possibility of transfer of DNA from one individual to another through intermediary items should always be considered.

Individuals may come into contact with a high quantity of objects within a short space of time and, as a result, the persistence of any *non-self DNA* acquired by the hands is subsequently affected by the activities performed and the duration since the initial transfer (van Oorschot et al., 2015; Szkuta et al., 2017).

One study aimed to measure whether a DNA source could be detected after multiple transfer events (Lehmann et al., 2013). Experiments were performed from cotton to cotton, and glass to glass, for wet and dried blood and 'touch DNA'. Transfer from wet blood gave a full DNA profile well beyond secondary transfer on both cotton and glass substrates. In fact, full profiles were obtained from the *sixth* substrate of both wet and dried blood. 'Touch' DNA produced partial DNA profiles up to the *fifth* substrate. A later study found that 'touch' DNA produced full DNA profiles up to the third substrate (Fonnelop et al., 2015).

A very interesting finding from the above study demonstrated quaternary transfer of DNA (ibid.). There were unknown fragments of DNA on some of the transfers investigated. These transfers were mediated by a particular individual and the DNA was from first, second and third substrates. These DNA fragments were later found to match the DNA of the girlfriend of the individual. The girlfriend had not visited the office where the sampling had been performed and had not been in contact with the individual for 10 hours prior to the experiment. The individual had also washed his hands before the experiments. One could surmise that this individual touched personal items such as clothing or cell phone once the experiment had commenced but this aspect was not investigated.

Another Norwegian study (Fonnelop et al., 2017) noted that indirect transfer may occur if two relevant individuals have both occupied the same area before the crime event. A participant in the transfer study in an office was found to have DNA on the front of her clean T-shirt, provided for the experiment, that corresponded to DNA from a person in the office who had been on leave for *2 months*, and who had not visited the office during that time. The modes of transfer could not be determined because the original deposit from the person on leave was not located.

It has been found that several factors may influence indirect transfer of DNA. These include the type and amount of the biological substance deposit, the nature of the primary and secondary (and further) substrate, the moisture content of the deposit and the type of contact between the surfaces (Fonnelop et al., 2015; Gosch and Courts, 2019). All these factors are those typically considered in the transfer of *trace material*, in general.

There is still insufficient knowledge regarding variables that may influence secondary or higher DNA trace transfer (van Oorschot et al., 2019). This lack of knowledge makes it difficult to assess alternative case scenarios conditioned on the evidence observed. Accordingly, caution is still advised when considering how DNA from different individuals may have been transferred to the object from which it was collected.

One paper (Helmus et al., 2016) stated that 'Every DNA transfer scenario one can imagine seems to be possible'. Gill (2016) also noted the 'imponderability' of the various speculative causes of DNA transfer.

3.2.2.3 Risk of Indirect Transfer – Social and Domestic Situations

Research to evaluate the risk of indirect transfer has not kept pace with technological developments and the increase in sensitivity of DNA typing techniques (Fonnelop et al., 2015, 2017). Research into indirect transfer of forensic DNA has until recently been limited and focused mainly on the handling of DNA-free items (Meakin and Jamieson, 2013; Meakin et al., 2015). DNA from unknown sources, such as background or prevalent DNA already existing in the environment, can be adventitiously transferred within the environment of the crime scene, with the potential to interfere with casework samples and further DNA profile interpretation such as number and type of contributors.

The following case that made headlines in numerous countries concerned trace DNA evidence for the conviction of two accused, the United States citizen Amanda Knox and her Italian boyfriend Raffaele Sollecito (Hellmann, 2011; Hanlon, 2011; Vecchiotti and Conti, 2011; Balding, 2013; Gill, 2014; Gill, 2016).

CASE 3.2: TRANSFER: SOCIAL CONTACT BETWEEN VICTIM AND ACCUSED

A British exchange student, 21-year-old Meredith Kercher, was stabbed to death in Perugia, Italy, in 2007. She was found deceased in her own bedroom in the apartment she shared with three other female students including Amanda Knox. She was found on the floor with stab wounds to her throat, sexually assaulted and some of her belongings had been stolen. The Ivory Coast born Rudy Guede was convicted in 2008 of murdering and sexually assaulting Kercher and sentenced to 30 years, reduced to 16 years on appeal in December 2009. The evidence against Guede appeared to be uncontroversial with DNA profiles matching his DNA profile on Kercher's body and clothing, and he had no legitimate access to the premises.

Amanda Knox and Raffaele Sollecito were also charged with the murder. They were convicted in 2009 but after they had spent 4 years in prison, an Appellate Court in 2011 acquitted them, accepting defence arguments on lack of evidence and potential contamination, and they were released. The acquittal was overturned by the Supreme Court of Cassation in 2013 and a retrial convicted both Knox and Sollecito again. There was a further appeal in 2015 called the Marasca-Bruno motivation during which both Knox and Sollecito were exonerated and a miscarriage of justice was officially confirmed.

Key evidence in the case against Knox and Sollecito was DNA obtained from a knife, found in the kitchen drawer at Sollecito's flat, and DNA from bra clasps from the deceased located at the crime scene. The knife allegedly had traces of DNA from Amanda Knox on the handle and of Meredith Kercher on the blade. The DNA alleged to come from Knox was not disputed (she regularly visited her boyfriend's flat) but the DNA profile alleged to come from Kercher was low level, not repeatable and could not be related to blood. It was not obvious why the knife was believed to be evidential (that is, the murder weapon). Police officers did not remove other knives (to test for background DNA) and questions were raised about handling and packaging. The suspects and victims knew each other and had access to each other's apartments.

The bra clasps were recovered from the scene 46 days after the crime in a context highly suggestive of environmental contamination. The bra of the victim had been cut and was supposedly collected at her feet. The bra clasp was a fragment of material with a deformed clasp that had been removed from the rest of the bra and originally observed under the deceased's body. The DNA profile from the bra clasp displayed a clear major/minor mixture with the major having come from Kercher. The minor component was alleged to have come from Sollecito. Y-STR profiling revealed a mixture of at least three males. Social media had leaked videos of the crime scene investigation showing scene examiners wearing dirty gloves. It was accepted that Sollecito may have contributed to the DNA but the question still remained as to how it was deposited. The defence proposed that an investigator may have inadvertently transferred DNA from Sollecito on the door handle of the bedroom (Sollecito agreed he touched the door handle to try to enter the room to check on the possible victim) to the bra clasp.

The final court could not rule out contamination for either the knife or the bra clasp. The court stated that low-level DNA precautions did not appear to have been applied for the knife DNA, and there was an erroneous interpretation of both the autosomal DNA and Y-STR profiles on the bra clasp.

An argument had been put forward that it was not enough for the defence to say that the DNA result was from contamination – the burden was on those claiming contamination to prove its origin. However, the Appeal Court held that the 'burden' was in showing the result was obtained using a procedure which guaranteed the integrity of the item, from the moment of collection to the moment of analysis. Once there was no proof that these precautions were taken, then it is not necessary to also prove the specific source of the contamination.

One Norwegian study simulated part of the above scenario by examining the transfer of trace DNA from a person's hand to a metal door handle, then to a piece of material through the gloves of an investigator (Fonnelop et al., 2015). It was found that the simulated tertiary transfer (person to door handle to glove to material) could produce the DNA of the original person handling the knob on the material of the bra clasp.

This case is illustrative of the prosecution inferring the association of an *activity* such as *stabbing* with a knife or *handling* a bra clasp, with a DNA profile that could not be sourced to a particular time or body fluid from known domestic associates. Furthermore, collection procedures were an issue in this case. The case is further discussed later in this book as it highlights many issues that may contribute to the misinterpretation of trace DNA evidence.

3.3 Speculative Recovery of Trace DNA

3.3.1 Recognition and Preservation

The key to obtaining and recovering sufficient and relevant DNA material from trace DNA samples is recognizing those items which may be suitable for such analysis. Another important key is to understand the appropriate sampling tools and method of sampling to achieve such a goal. This is no different to the sampling and testing rationale for any forensic trace or deposit on an item (Taupin and Cwiklik, 2010). Often, the desire of crime scene personnel to collect any DNA possible at the scene may lead to a 'swab-a-thon' at crime scenes (Butler, 2015) with too many samples collected of dubious value, the wrong area sampled or too little/too much sampled. Forensic laboratories may have a more restricted approach but, nevertheless, the fact remains that it is unknown if DNA capable of producing a profile is present in the sample – until the visualization of the DNA profile after amplification and separations of fragments.

The persistence and preservation of DNA also depends on the quality and quantity. Trace DNA samples may suffer from the problem of low-quantity DNA and/or DNA that is degraded in some way, or otherwise unable to be extracted properly from the material on which it is deposited. Thus recovery techniques need to be optimal for the surface to sample or collect potential DNA material.

Recovery and subsequent profiling of DNA material that is not present as a visible stain is now considered routine and is referred to as 'trace DNA' analysis (Bond and Weart, 2017). There are a number of factors that may affect the potential to recover trace DNA from a particular surface and this should be considered in any sampling and testing rationale. A single optimal sampling technique does not exist (Gosch and Courts, 2019). Even the individual performing the sampling process may influence not only the optimum of the trace examined (ibid.) but also may introduce the potential for contamination. Chapter 7 discusses contamination at the crime scene and laboratory.

Other authors compared the effectiveness of trace DNA profiling on evidential exhibits that had potentially been touched, between a jurisdiction in the United States and a jurisdiction in the United Kingdom (Bond and Weart, 2017). Results showed greater numbers of USA-firearm-related items submitted for analysis compared with the United Kingdom where greater numbers were submitted from burglary or vehicle offences. The authors emphasized that careful selection of both items and sampling techniques is crucial to obtaining relevant and informative DNA results (ibid.).

The types of exhibits examined for 'trace DNA' deposits will depend on the particular jurisdiction, the laws involved and other social aspects. For example, due to the strictly regulated laws of gun ownership in the United Kingdom and countries like Australia, the prevalence of gun-related crime is much lower than the United States. However, knives and other implements may be examined more as evidential exhibits in these jurisdictions.

When a potential deposit of trace DNA is to be collected, the first question the examiner needs to ask is 'What is the potential area to sample and how should I sample it?' There have been many studies in order to address this question. Exhibits may be swabbed or tape lifted based on assumptions about where the DNA material is located if there is no visible deposit (van Oorschot et al., 2010).

An interesting observation was made in a study on the prevalence of DNA of regular occupants in motor vehicles, driven for a week both with and without a front seat passenger (De Wolff et al., 2021). One vehicle had a lower-than-average amount of DNA yet was used in a similar fashion to the other vehicles. It was postulated that the reason was the soiling of the surfaces of the vehicle with the study participant saying 'often dogs in the back … occasionally on the front seats … dog may press horn on the steering wheel'. Sampling efficiency was potentially reduced due to the rapid saturation of the tape used for sampling (ibid.). Thus, the prevalence of *other* trace material (such as dog hair) may impact on the recovery process of sufficient DNA for testing. It may also be relevant to consider *other* trace material in the context of the case. Chapter 6 further discusses DNA in motor vehicles.

The above paper (ibid.) also noted that 95% of samples from the motor vehicles yielded a DNA profile, a higher success rate compared to samples from vehicles in casework such as one reported study where 29% of samples from vehicles yielded an interpretable profile (Mapes et al., 2016). It was believed by De Wolff and authors in their paper that such differences are likely caused by differences in experimental and casework conditions (e.g., persistence of DNA negatively impacted by an extended period between deposition and sampling, packaging and storage) as well as DNA analysis systems with different sensitivity and reporting guidelines (de Wolff et al., 2021). It was also noted that other differences were reported for other sample types such as cable ties (Steensma et al., 2017), discussed further below.

It is *imperative* for all those interacting with exhibits potentially containing biological material to have an awareness of DNA transfer possibilities after the criminal activity occurred, in order to limit or mitigate the risk of contamination or loss of DNA. This is no different to the main priority of any scientific examination – preserving the integrity of the exhibit before analysis and performing that analysis with a sampling and testing rationale conforming to the scientific method. If DNA analysis is required at a later date then DNA contamination mitigation procedures and awareness should be in place. When 'cold cases', or those where trace DNA transfer was not a priority in collection or analysis, are investigated then the necessary caution should be applied. Otherwise serious consequences may occur such as demonstrated in Case 1.3 (Farah Jama), Case 2.5 (Kevin Brown), Case 3.1 (Phantom of Heilbronn) and Case 3.2 (Meredith Kercher).

3.3.2 Relocation of DNA

The 'mobility' of DNA has been demonstrated in numerous studies. Distribution of DNA traces by wiping or scrubbing an object, with water or other cleaning fluid, has been shown (Helmus et al., 2018). This may be an issue when using presumptive screening tests for blood, semen or saliva at the crime scene or laboratory, when reagents or water are used in the testing strategy.

This author uses the term 'relocation of DNA' to describe the relocation of a DNA deposit to another area during forensic collection, handling, sampling and testing. Essentially this is transfer of DNA through methods used in forensic analysis. This includes searching for all forensic material, not just biological material.

3.3.2.1 Relocation of DNA through Handling and Testing (before Biological Analysis)

An Australian study found that significant quantities of DNA are frequently: (a) transferred from an exhibit to the inside of its packaging and (b) transferred from its area of initial deposit *to other areas of the same exhibit and/or to other exhibits within the same package* (Goray et al., 2012; this book author's italics). DNA could be transferred from the deposit area to either other parts of the item or to the bag itself – and usually to both. Loose bags can, in certain situations, permit more transfer, allowing greater movement of an item within the bag, thus providing more friction between surfaces and hence opportunities for transfer. DNA was lost to the inside of containers holding bloodied knives and of concern there was redistribution of the DNA-containing material from the tip of the blade to other areas of the knife, including the handle.

Another experiment in the same study examined blood on underpants. Samples of 100 micro litres of blood taken from two individuals were deposited on the front middle of each garment, respectively. Prior to sample deposit, a sterile transparency was placed inside each piece of underwear to prevent leakage. The deposits were allowed to dry for 24 hours. Two underwear items, one from each depositor, were placed into brown medium-sized exhibit bags and handled as per collection and transport from a crime scene. DNA from the deposit area was observed on each of the three non-deposit areas of the same underwear in 100% of garments: front-outside, rear-outside and inside. DNA derived from the deposit was also observed on each of the three non-deposit areas of the *other* garment that was within the package in all garments: front-outside, rear-outside and inside. DNA from both deposits was also retrieved from inside the packaging in 60% of the brown paper bags (ibid.).

The underwear experiment generated DNA profiles from all garment surfaces. The profiles generated from the deposit sides (front-outside) as well as the rear-outside contained significant proportions of alleles derived from both the depositor and the non-depositor. In some instances, the major profile in the mixture was from a non-depositor. The inside of the underwear contained the largest proportion of transferred DNA compared to other areas. This finding could be due to the random positioning and placement of the underwear in the bag, whereby DNA could be transferred.

Cigarette butts from two different users, and un-smoked cigarettes cut to butt size, were placed in A5-sized yellow paper envelopes and transported as per police conveyance to a laboratory (ibid.). It was found in this study that over 85% of un-smoked butts packaged with the smoked butts had DNA profiles on them.

These findings, made in 2012 using less-sensitive techniques than those used today, highlight the need to deal with issues inherent in the collection and packaging of exhibits for forensic DNA analysis. It also calls into question the relevance of the location of DNA on exhibit surfaces and interpretation made from the same.

Further research is necessary to explore these findings.

3.3.2.2 DNA Relocation during Forensic Testing Prior to Biological Examination

Fingerprint brushes and reagents may be used to help visualize potential fingerprints at the scene or in the fingerprint laboratory, before submitting the items for DNA sampling. The potential for the movement of DNA and/or contamination of the original deposit needs to be considered. A study in 2005 found that fingerprint brushes could potentially collect and transfer DNA and the same brush could powder different items of evidence within and between crime scenes (van Oorschot et al., 2005).

The dusting of latent prints may dislodge cellular debris; that debris may adhere to the brush. This brush can then potentially be used on another item where it also may transfer or dislodge cellular debris. The study recommended that when fingerprinting, biological evidence should be avoided if possible, avoid powdering areas that may be sampled for DNA analysis, use separate fingerprint brushes and prepare and use separate aliquots of powder.

It has later been noted that secondary transfer of DNA from exhibit to exhibit can occur via fingerprint brushes and is enhanced using low-level DNA analysis (Bolivar et al., 2015).

Superglue chambers are also used in the fingerprint laboratory to develop latent fingerprints for easier visualization. Superglue is heated and turned to vapour in a controlled chamber with the evidence item. Because the chamber contains vapourized particles, the movement of particulates within it is possible. A study (Gibb et al., 2012) showed that DNA has the potential to accumulate and transfer within the chamber. Standards were suggested in this paper to prevent such DNA contamination.

Non-invasive detection systems, that may highlight the areas to examine, are desirable. These are alternative light sources that are not specific for any biological matter but may indicate a deposit of biological matter on the surface examined. Such instruments as the Polilight (van Oorschot et al., 2010) may be useful in locating potential areas of deposit. Swabbing or tape-lifting an area smaller than the actual deposit may mean that insufficient DNA is collected. Conversely, sampling an area greater than the actual deposit may complicate any interpretation.

The important step of collecting an exhibit at a crime scene, medical examination or autopsy room – and subsequent packaging and handling and transport to the laboratory – is critical for any sampling and testing rationale. Inadvertent transfer of DNA in the medical room has been described in Case 1.3 and inadvertent transfer at the crime scene in Case 3.2.

3.3.2.3 Relocation of DNA through Biological Testing

'Luminol' screening testing is a forensic biology presumptive tool used at crime scenes for non-visible traces of blood. It is sprayed in a liquid form over the potential deposit and, depending on the amount sprayed, drops or even pools of the reagent may form over the surface. There have been studies performed on the effect of luminol on the subsequent detection of DNA; however, to this author's knowledge there has been no discussion regarding movement of DNA traces on application of luminol.

A trial which subsequently had many appeals, including two applications to the High Court of Australia, had luminol used in the original examination for presumptive testing for blood on the deck of a yacht. A DNA profile was obtained that could not be related to a biological fluid – trace DNA. Multiple debates revolved around direct and secondary DNA transfer of the deposit (*Neill-Fraser v. State of Tasmania*, 2019, 2021).

CASE 3.3 MURDER ON A YACHT

A 56-year-old woman in the southernmost island State of Australia, Tasmania, was convicted at trial by jury in 2010 of murdering her 68-year-old partner Bob Chappell in January 2009 on their yacht 'Four Winds'. The case consisted entirely of circumstantial evidence and the body has never been found. It was alleged that Sue Neill-Fraser had gone to the yacht where her partner was alone, killed him and winched his body into a dinghy and then disposed of him. She appealed against her conviction in 2012 but it was upheld, although her sentence was reduced to 23 years with a non-parole period of 13 years. She sought special leave to appeal to the High Court but that request was refused. Amendments to the criminal legislation meant that she could appeal to the Court of Criminal Appeal if there was 'fresh and compelling evidence'.

Neill-Fraser thus appealed due to the evidence of DNA found on the bow of the deck of the yacht. Luminol screening had been conducted on the yacht over a period of days in 2009, and positive results had been obtained from inside the dinghy and on various areas of the yacht including inside the cockpit (which produced a DNA profile corresponding to Bob Chappell). There was also a luminol positive area from the starboard walkway of the deck about 210 by 260 millimetres in dimension. This area was sampled with a swab which yielded a single-source DNA profile, said to match through a database hit to a homeless teenager in 2010. During the trial the teenager, Ms V, denied being on the yacht.

Prior to the DNA being sampled some 21 people – including workers, investigators, firemen and family members – were allowed access to the yacht. It was considered by the courts as reasonable to infer that most of these people had walked across the relevant area on the deck more than once. Author note – consequently the crime scene had already been compromised prior to forensic examination.

The prosecution argued that the presence of the DNA on the deck was a 'red herring' and that the DNA could have come aboard the yacht by way of secondary transfer. In particular it was argued: 'But it could have been put there at any time before the DNA swab was taken by anyone who had acquired some trace on their footwear at any place and then maybe got in the car, driven down and got out and onto the boat and transferred it. All those things are logically possible, all things go to explain this finding' (*Neill-Fraser v. Tasmania*, 2021). Author opinion – the explanation of secondary transfer contradicts the 'usual' prosecution theory of the assumption of direct transfer if a DNA deposit is found.

The only expert evidence concerning secondary transfer of DNA at the trial was provided by a State laboratory forensic scientist. The scientist said, upon questioning, that secondary transfer was a possible explanation for the presence of the DNA on the yacht, but, during cross-examination, stated that he was not prepared to express an opinion one way or the other as to whether it was more likely that the deposit had taken place as a result of primary or secondary transfer. With respect to the suggestion that the deposit had been transferred onto the yacht on the bottom of someone's shoe (a prosecution theory) the forensic scientist said that he had not seen or experienced such a scenario before.

During the second appeal (leave to appeal), a forensic scientist from another State laboratory was prepared to say that although it depended on the surrounding circumstances, the nature of the DNA profile was not typical of secondary transfer (*Neill-Fraser v. Tasmania*, 2019). The scientist also said the DNA evidence was unlikely to be more than 2 days old. Further, the type of DNA profile was more likely to have come from body fluids (blood, saliva, etc.) than a simple skin contact/touching event. The scientist also discussed the rarity of secondary transfer having occurred by someone stepping into a biological substance containing DNA and transferring it to another location on the sole of their shoe, and the circumstances required for such a transfer to have occurred.

The second appeal (*Neill-Fraser v. Tasmania*, 2021) was delayed due to Covid-19. The now young woman Ms V whose DNA was agreed to be on the yacht (including on a TV interview program '60 minutes'), said she had been on the yacht with three young men, and at least one of the men hit the owner of the yacht. At the appeal hearing the young woman initially stated she saw a lot of blood, panicked and vomited on the deck – but then under cross-examination denied being on the yacht.

The 2021 appeal ruling was a majority decision of two to one, stating that the new evidence did not meet the required threshold of 'fresh and compelling'. The ruling noted it was important to bear in mind that the information regarding the nature of the DNA profile was highly relevant to the defence and the hypothesis that Ms V and/or her associates were responsible for the death of Mr. Chappell. Questions such as how long the DNA might have been there, whether it was degraded and how it could have got there were highly pertinent to the defence case.

On 12 August 2022, the High Court of Australia refused special leave to appeal the second appeal (James and Brown, 2022). Sue Neill-Fraser, now 68 years, was released on parole on 4 October 2022.

During the original trial the scientist who performed the luminol examination on the yacht had stated in their report that the luminol positive area on the deck of interest had 'possible drops, negative with Hemastix screening test for blood'. In testimony it was explained that it can be difficult to determine the difference with a stain formed from a drop of the deposit or a drop from the luminol bottle. It was possible that there was a trace of blood mixed in with other body fluid because luminol is very sensitive, but it was also possible that the luminol positive material had nothing to do with the DNA result.

The appeal ruling noted that the questions asked of both scientists strayed beyond the limits of science and into the areas of the chance and likelihood of two causal mechanisms as explanations for the DNA deposit. The 'new' evidence was not highly probative with respect to the question of the hypothesis advanced by the defence that someone other than the appellant was responsible for the death of Mr. Chappell (*Neill-Fraser v. Tasmania*, 2021).

Note that there appeared to be no visible deposits on the deck but a luminol examination was performed and an area was collected by swabbing a chemiluminescent region. The reagent fluid may have affected the original deposit area in shape and dimension. Luminol should be used as a screening test only. No biological fluid origin was determined – there was a presumptive test for blood (Hemastix) which was negative but no further testing was performed. This trace DNA was a full single-source DNA profile but neither quantity nor quality can determine whether the deposit is from 'touching' or other means. Nor can a DNA profile be 'aged' as to deposition, specifically regarding days of exposure and considering the effect of a chemical reagent such as luminol.

Debating primary or secondary transfer delves into explanations of the DNA results, not an evaluation of the result in light of different scenarios. Commenting on the probability of one scenario versus another is not the domain of the scientist as the court appeal points out (this is transposition of the conditional). Chapter 2 discusses the issue that when answering questions such as, 'Is it possible there was secondary transfer?' then this is giving an opinion on the alleged activities implying secondary transfer.

Experiments regarding the transfer of DNA onto a surface via walking were not examined in the court –such as an example of a secondary transfer scenario as proposed by the prosecutor to account for the DNA from Ms V. Explaining away an inconvenient DNA result by contamination (or inadvertent transfer) should throw other forensic results into doubt, including others in the jurisdiction.

Whilst released, Sue Neill-Fraser still agitates for her exoneration. This author questions whether experts would be prepared to debate the DNA deposit of Ms V in the light of activity-level evaluations. As an example, one hypothesis could be that 'the DNA evidence of Ms V occurred through vomiting on the deck' and the alternative 'the DNA evidence of Ms V was trod onto the deck of the yacht on a shoe'. Both scenarios should also have a luminol spray at the end, thus evaluating the DNA obtained in the light of two alternative hypotheses/activities.

It should be acknowledged that in 'cold cases', or even recent cases, proper contamination mitigation procedures need to be in place from the receipt of the item to the final reporting of the DNA result. Further, there should be an acknowledgement of any previous testing performed on the item. Case 2.5 of Kevin Brown illustrates the problems of cold case re-examination without proper recognition of previous testing.

The practice of screening (presumptive testing) of clothing items or bedding for the presence of seminal acid phosphatase in semen requires the consideration of DNA-free surfaces on which the exhibit is laid and DNA-free tools prior to wetting with the reagent, which may not have been true in the past. The surface required (then and now) may have needed to be a large bench sufficient in area to encompass sheets or quilts but also smaller items sprayed with the reagent such as under garments including underpants and brassieres. The potential of DNA transfer from previously examined exhibits is unknown.

Once wet with the reagent for semen screening, the clothing may have been dried so that it could be packaged. Wet clothing items from the process were placed on 'drying racks' in the laboratory akin to clothing airing stands. There was a potential for transfer of DNA from clothing item to clothing item on these drying racks or from rack to item.

Any potential DNA transfer events post examination need to include the considerations of the examination of the exhibit before any subsequent sampling, especially in cold cases.

3.3.3 Collecting Trace DNA Deposits – Sampling and Testing Rationale

Sampling techniques used to collect a deposit may include swabbing, tape-lifting and cutting out of material. DNA recovery is dependent on the efficiency of the sampling techniques. However, there is currently a paucity of meaningful sampling success rate data available for comparison (van Oorschot et al., 2019).

There are a number of methods for collecting trace DNA which include swabbing (sometimes multiple swabbing with wet and dry swabs), scraping, adhesive tape, cutting out samples and wet vacuuming. It is imperative that all these collection tools are DNA free before sampling, and not just sterile, as exemplified by the Phantom of Heilbronn (Case 3.1).

Trace DNA samples collected by swabs is a traditional tool. Swabbing an area requires a moistened swab to traverse the whole target area multiple times with some pressure and rotation of the swab so that the full surface area of the swab can contribute to the collection. Still, a moist cotton swab may not pick up all available DNA on the surface. It is common practice to perform a double swabbing technique which has a moistened first swab and a second dry swab to collect the remaining DNA (van Oorschot et al., 2010).

A paper on 'wearer DNA' of clothing (Breathnach et al., 2016) observed that the variability of different laboratories in detecting the wearer of a garment may be attributable to varying sampling methods and analysis used by those laboratories. This finding can be extrapolated to items in general.

This author worked in the United Kingdom during the 2000s when adhesive tape-lifting techniques were used on all absorbant surfaces, such as fabrics in clothing, while swabbing techniques were used on non-absorbant surfaces such as knives, for exhibits examined in the laboratory and at crime scenes. The United Kingdom crime scenes were attended by dedicated CSI units (Bond and Weart, 2017). This paper noted that in the United States there is a relatively high reassignment of officers and appropriate training is not necessarily passed down to newer members of the police force.

A demonstration of the above can be seen in the greater success in obtaining DNA profiles from material such as vehicle headrests in the United Kingdom. The use of adhesive tapes to recover trace DNA from fabrics in the United Kingdom is more effective at recovering trace DNA from absorbant materials than swabbing techniques (Hansson et al., 2009). Adhesive tapes also reduce the co-sampling of inhibitors present in the fabric such as clothing dyes compared with wet-swabbing techniques (Barash et al., 2010; Verdon et al., 2014).

Detection rates can also be dependent on the analysis and interpretation methods used in a particular laboratory. One study compared four laboratories in their interpretation of DNA deposited on cable ties (Steensma et al., 2017). Results of the study showed significant differences in the amounts of DNA recovered by the four laboratories. The reportable DNA profiles showed differences in the number of mixtures versus single-source profiles obtained. It was also demonstrated that packaging, transport and time delay before sampling of the items affected the quantity and type of DNA profiles obtained.

There have been attempts to recover DNA directly without the extraction stage in the aim to maximize recovery from low template/degraded DNA (Templeton et al., 2015). Some surfaces provide difficulty for recovery of DNA such as brass cartridge cases. An optimized swabbing technique and 'direct PCR' was used on touched metal cartridge cases, glass and tape as part of a mock case study. Good quality DNA profiles were obtained from all surfaces, and short exposure to UV light did not appear to adversely affect DNA recovery. However substrates left outside and exposed to environmental conditions for up to 8 days may reduce success rate.

There have been recent alternatives to sampling when there are difficult porous surfaces such as the 'M-Vac' system, which has been described as a wet-vacuum alternative and a 'mini-hurricane of sterile solution combined with vacuum pressure' (Taylor, 2020). Wet vacuuming after swabbing yielded an average of ten times more DNA compared to the initial wet swabbing. The results demonstrated that considerable DNA remained in or on the substrates after wet swab collection. This technique is more resource intensive and may not be suitable for many cases.

Although swabbing and taping seem relatively straightforward techniques there have been instances of how easy it is to get them wrong (van Oorschot et al., 2010). Inadequate training, combined with the absence of competency testing and ongoing monitoring of an individual's techniques, could drastically limit the success rates.

Investigating laboratories have different methodologies and analytical tools. Regarding DNA profiles, there have been continuing efforts to understand inconsistencies of results between laboratories using the same or different methods on the same sample and recommending practices that can be universally applied to reduce their impact. Guidelines and standards have been produced in the United Kingdom (issued by the Forensic Regulator) and the United States (issued by the Academy Standards Board of the American Academy of Forensic Sciences) to provide a uniform guide for laboratory practice.

Even in co-operative experiments, large differences in some results have been observed from the same experiments. An interlaboratory study across four laboratories in different countries found that DNA profile composition appeared to be laboratory dependent when performing mock 'hugging' experiments with a 'wearer' and a 'hugger' (Szkuta et al., 2019). One laboratory generated mostly single-source DNA profiles corresponding to the wearer whereas the other three tended to produce mixtures where the wearer was the major contributor.

3.4 Trace DNA Quantities

It is not correct to assume that potentially 'touched' objects, or any trace DNA material in question, contain only low amounts of DNA (van Oorschot et al., 2010; Meakin and Jamieson, 2013; Atkinson et al., 2022). Depending on the nature, frequency and duration of the contact, tens or thousands of nanograms (a nanogram is one thousand millionth of a gram) of DNA may be present. One DNA cell contains about 6 picograms (a picogram is one thousandth of a nanogram) of DNA. Laboratories process samples even when no DNA is quantified as a DNA profile may still be produced (author experience in United Kingdom laboratories).

When little or no DNA is quantified it is believed that with random sampling (stochastic) effects, the low amounts of template DNA are only an indication of the concentration (van Oorschot et al., 2010). Consequently the amount of DNA detected using traditional methods and the results obtained may not be a real reflection of the amount of cells analyzed. This is an important, but often overlooked, concept.

There are difficulties with improving the amplification and typing of trace DNA so that sufficient DNA is detected for a result. None of the commercial kits are validated by their manufacturer for very low template amounts (van Oorschot et al., 2010). This is true as of this writing. The lack of validation is partly due to the innate variability that exists with trace DNA samples and their analysis. Any trace DNA sample may be subject to the inherent variation which comes with operating at such low levels.

3.5 No Relevant Deposit Detected

A common saying in medical science research is 'absence of evidence is not evidence of absence'. It may be used in scientific experiments if there is a hypothesis of an effect, but such effect is not observed. However, this may be due to a lack of appropriate testing, poor experimental design or limitations of the testing. This of course applies to forensic science research.

One cannot infer the absence of contact from the absence of a trace because contact may occur without a trace being found (Thompson and Scurich, 2018).

A recent study (Atkinson et al., 2022) on handled items in an experiment found that there were many instances where the DNA of an individual, known to have contacted the item, was not observed as a component of the corresponding DNA profile. The authors note that this 'reaffirms the oft-quoted adage that the absence of an individual's DNA within a profile cannot be used to draw the conclusion that this person did not handle or make contact with the object' (ibid.).

'Negative' forensic evidence can be defined as the failure to find a trace after looking for it, but sometimes the absence of evidence can be highly probative (Thompson and Scurich, 2018). These authors quoted an Arthur Conan Doyle story of 'The curious incident of the dog in the night time' where the inspector in the case stated to Sherlock Holmes that the dog did nothing and the reply was 'that was the curious incident'. The perpetrator in the story had acted by night and the dog was likely to have barked upon encountering a stranger. Thus the failure of the dog to bark was highly probative and the key to solving the case since the perpetrator was no stranger to the dog (ibid.).

Regarding forensic biology cases, failure to find semen traces in a vaginal sample a month after alleged sexual intercourse provides no evidence in a sexual offence case due to the time lapse. Thus the absence of evidence is uninformative – and indeed, a sample would not be taken by medical personnel.

However, in cases where the probability of detection falls somewhere between 100% and 0% the probative value cannot either be all or nothing, definitive proof or worthless (Thompson and Scurich, 2018). A likelihood ratio for negative evidence specifies the relative likelihood of a negative finding about alternative hypotheses regarding some activity (ibid.). Such concepts may be difficult to understand due to inferential reasoning, and it is contrary to the principle of falsification of hypotheses (this author note). In order to assess the value of negative evidence, forensic scientists need to think about the probability of a negative finding under the relevant hypotheses and need empirical data on the probability of detecting traces of various kinds under various circumstances (ibid.).

A lengthy case from Australia demonstrates issues in conveying information to the legal system and how the absence of DNA evidence in a case can become a bone of contention over numerous years and court proceedings (*R v. Drummond* [2013], [2015]). It has also been used to advocate for activity-level propositions in forensic science papers (Taylor, 2016; Gill et al., 2020) and is relevant to trace DNA.

CASE 3.4 *R V. DRUMMOND*

The accused Adrian Drummond was charged with the offence of attempted kidnapping. A 15-year-old girl was walking on the footpath in South Australia in 2010 when she stated that a slow – moving car drove towards her. She said she became worried and entered the registration number of the vehicle into her phone. A male alighted from the vehicle and grabbed her arm to get her into the car. The complainant broke free after grappling with the male. The description of the vehicle was not disputed but the complainant picked another person out from the identification procedure. During March 2012 the accused was convicted by a jury verdict following a trial and was sentenced to 5 years and 3 months imprisonment. He appealed against the conviction to the Court of Criminal Appeal. The appeal was dismissed. He was then refused special leave to appeal to the High Court of Australia. Drummond then sought permission to appeal for the second time on the ground that there existed fresh and compelling evidence that should, in the interests of justice, be considered on appeal. The second appeal was successful whilst Drummond was on parole after serving 2 years and 3 months. The prosecution dropped charges after a retrial was ordered (McGregor, 2015).

A forensic scientist from the state laboratory gave evidence at the original trial of attempts to recover traces of DNA on the clothing tops of both the complainant and Drummond. No trace of Drummond's DNA was found on the complainant's clothing, nor was any trace of the complainant's DNA found on Drummond's clothing. There was a mixture DNA profile of three contributors including the complainant on her top and a mixture DNA profile of two contributors including the accused on his top garment.

The forensic scientist's opinion during the trial that no usable DNA material from the accused was located did not necessarily mean that there was no contact, although they stated that it was a possibility. DNA material might not be found on an object even where a person has come in contact with the surface of that object. This included reference to a study in the local forensic science laboratory that showed that only about 10% of samples actually provide any useful information. The forensic scientist presented another study at the second appeal that showed clothing yielded a DNA result in about 90% of samples. Note that this discussion delves into activity levels of DNA-TPPR which sub-source levels of DNA cannot address.

On the first appeal the import of the 10% figure was not recognized whilst the second appeal found the figure was misleading. Two judges allowed the second appeal but the third judge disagreed. The 10% figure had come from the Appendix to the forensic laboratory report that stated:

> '**Contact DNA** – Contact DNA refers to biological material left on an object through a contact transfer (such as touching, handling or wearing) and not via deposition of a biological fluid (such as blood, semen or saliva). Contact DNA samples usually contain only small quantities of DNA and therefore analyses often do not give an informative profile. (Laboratory) studies have found that of the majority of commonly submitted contact DNA sample types, only about 10% yield an informative DNA profile'
>
> **(Taylor, 2016).**

One second appeal judge noted that when the testifying scientist stated '*usable DNA is recovered in only 10 per cent of DNA testing*' did not mean usable for the purposes of the particular case but rather usable, or suitable, for the purpose of uploading a DNA profile into a computerized system or database. As quoted in the appeal court ruling of Sue Neill-Fraser Case 3.3, the judge noted in the Drummond case: 'I have no doubt that the jurors in the present case would have gained the firm impression that they were being told that they could confidently apply such evidence to the case before them, and that they could do so on the basis that the statistics referred to … were indeed logically applicable to the present case'. As Peek J. stated, the appeal of Drummond (No 2) was highly unusual in that it involved the giving of evidence by a prosecution expert witness that subsequently had been demonstrated to be incorrect.

The two appeal judges agreed that reliance on an average success rate over all types of articles and surfaces subjected to DNA testing rather than the figures referable to testing of clothing was a serious flaw in the forensic scientist's evidence.

Adrian Drummond consistently denied his involvement in the offence, was of previous good character and the imprisonment seriously impacted him and his family. The State government subsequently denied him compensation.

The Drummond case has been used in the forensic science literature promoting the use of 'activity' level propositions due to the courtroom discussions regarding absence of DNA and the alleged scenario of touching, which 'source' level propositions cannot address. As an example it has been stated that the finding that no DNA of a person of interest has been detected *does* have evidential value (Taylor, 2016). This particular paper evaluates the probability of the DNA results given the competing propositions and the information, in the form of a likelihood ratio. This likelihood ratio devolved to 'the DNA findings are in the order of 2 to 4 times more likely to have been obtained if the defense proposition had occurred rather than the prosecution proposition. This provides slight (or weak or limited) support to the proposition that the complainant had no contact with the accused compared to the complainant and accused struggled' (ibid). This relates to activity-level propositions. It would be interesting to know any judge or jury acknowledgement of what 'slight in favor of the defense' means in context of the case.

A guidance paper for forensic genetics has a section on 'absence of evidence' in relation to the absence of finding trace DNA or a DNA profile relevant to the person of interest, and notes the Taylor paper regarding the Drummond case (Gill et al., 2020). Consideration 3 of the paper states 'absence of biological evidence from a person of interest (POI) where there is an expectation of an observation under prosecution's proposition will generally support the competing defense proposition'.

The absence of DNA will rely on probabilities of transfer, persistence and recovery of DNA given the proposed mechanisms put forward by prosecution and defence. However, in this author's opinion, the issues with the absence of DNA could largely have been averted if the judge and jury had understood the correct meaning of the study put forward by the laboratory scientist in the trial. It was up to the laboratory and the reporting scientist to convey the true meaning of the study regarding 'usable DNA profiles' meaning uploading to a DNA database for searching and not comparison to a suspect reference sample. Indeed, the laboratory report should also have clarified the study purpose.

3.6 Context of Trace DNA

Items collected from crime scenes may not be free from DNA before coming into contact with a purported offender. A level of 'background DNA' may be present on the item, especially if it is regularly used such as an implement or clothing. DNA material from cells is ubiquitous material and can be found for reasons unconnected to a crime event. Thus, it is reasonable to assume that most surfaces and items will have some DNA on them. 'Background DNA' is discussed further in Chapter 4.

It is also important to be aware of crime scene collection practices and collection of the exhibit by police not trained as crime scene examiners. If a person of interest lives in the premises, it would be strange to examine an exhibit collected from those premises for their potential DNA. Collection and handling of exhibits prior to submission at the forensic laboratory must be rigorous when dealing with trace DNA. This subject is discussed in Chapter 7.

Early studies on finding DNA, that may not be from the offender, had focused on the role of contamination during examination of deposits in the laboratory. Material containing non-visible DNA could accumulate on tools and surfaces without the examiner being aware (Poy and van Oorschot, 2006). Implementation of environmental DNA monitoring programs were advocated as a useful means of identifying high-risk contamination surfaces, tools, procedures and practices and recommendations changed when more sensitive techniques were employed (Ballantyne et al., 2013).

Forensic science findings acquire meaning in context. One view is that scientists who produce such findings should be isolated from all information to avoid bias. However, forensic scientists need to separate impacts of biasing information that can be avoided and the contextual information that is vital to give meaning to the findings. The ethical onus on the scientist is to retain competence, which includes keeping up to date with the impact context has on findings.

The scientist needs to be transparent in their communication and capable of articulating the way in which they arrived at their conclusions, including the utilization of contextual information to frame their interpretation.

There are specific considerations when interpreting 'trace' DNA, or unspecified cellular DNA material, in context. More than a decade ago it was described that every trace DNA profile should be interpreted in the context of possible contamination (van Oorschot et al., 2010). A mixed DNA profile may contain background DNA, crime-related DNA and post-crime contamination, and it may be difficult to identify the relevant profile. The increase in cold case investigations using DNA profiling increases the risk of detecting trace DNA samples which may not have been collected, stored or examined with trace DNA sensitivities in mind (ibid.).

References

Atkinson, K., Arsenault, H., Taylor, C., et al., 2022, Transfer and persistence of DNA on items routinely encountered in forensic casework following habitual and short-duration one time use, *Forensic Science International: Genetics*, 60, 1–14.

Balding, D., 2013, Evaluation of mixed source low template DNA profiles in forensic science, *Proceedings National Academy Sciences*, 110, 30, 12241–12246.

Ballantyne, K., Poy, A., and van Oorschot, R., 2013, Environmental DNA monitoring: Beware of the transition to more sensitive typing methodologies, *Australian Journal of Forensic Sciences*, 45, 3, 323–340.

Barash, M., Reshef, A., and Brenner, P., 2010, The use of adhesive tape for recovery of DNA from crime scene items, *Journal of Forensic Sciences*, 55, 1058–1064.

Bolivar, P., Tracey, M., and McCord, B., 2015, Assessing the risk of secondary transfer via fingerprint brush contamination using enhanced sensitivity DNA analysis methods, *Journal of Forensic Sciences*, 61, 1, 204–211.

Bond, J. and Weart, J., 2017, The effectiveness of trace DNA profiling – A comparison between a U.S. and a U.K. law enforcement jurisdiction, *Journal of Forensic Sciences*, 62, 3, 753–780.

Breathnach, M., Williams, L., McKenna, L., et al., 2016, Probability of detection of DNA deposited by habitual wearer and/or the second individual who touched the garment, *Forensic Science International: Genetics*, 20, 53–60.

Butler, J., 2015, *Advanced topics in forensic DNA typing: Interpretation*, Academic Press, Elsevier, Oxford and San Diego.

Butler, J. and Willis, S., 2020, Interpol review of forensic biology and forensic DNA typing 2016–2019, *Forensic Science International: Synergy*, 2, 352–367.

De Wolff, T., Aarts, L., van den Berge, M., et al., 2021, Prevalence of DNA of regular occupants in vehicles, *Forensic Science International*, 320, 110713.

Diehl, J. and Juttner, J., 2009, *Suddenly the DNA was no longer there*, March 26, available online at http://www.spiegel.de

Fonnelop, A., Egeland, T., and Gill, P., 2015, Secondary and subsequent DNA transfer during criminal investigation, *Forensic Science International: Genetics*, 17, 135–162.

Fonnelop, A., Ramse, M., Egeland, T., and Gill, P., 2017, The implications of shedder status and background DNA on direct and secondary transfer in an attack scenario, *Forensic Science International: Genetics*, 29, 48–60.

Gibb, C., Gutowski, S., and van Oorschot, R., 2012, Assessment of the possibility of DNA accumulation and transfer in a superglue chamber: A preliminary study, *Journal Forensic Identification*, 62, 409–424.

Gill, P., 2014, *Misleading DNA evidence: Reasons for miscarriage of justice*, Academic Press Elsevier, London and New York.

Gill, P., 2016, Analysis and implications of the miscarriages of justice of Amanda Knox and Raffaele Sollecito, *Forensic Science International: Genetics*, 23, 9–18.

Gill, P., Hicks, T., Butler, J., et al., 2020, DNA commission of the International society for forensic genetics: Assessing the value of forensic biological evidence – Guidelines highlighting the importance of propositions. Part II: Evaluation of biological traces considering activity level propositions, *Forensic Science International: Genetics*, 44, 1–13.

Goray, M., Eken, E., Mitchell, J., et al., 2010, Secondary DNA transfer of biological substances under varying test conditions, *Forensic Science International: Genetics*, 4, 62–67.

Goray, M., Van Oorschot, R., and Mitchell, J., 2012, DNA transfer within forensic exhibit packaging: Potential for DNA loss and relocation, *Forensic Science International: Genetics*, 6, 158–166.

Goray, M. and van Oorschot, R., 2015, The complexities of DNA transfer during a social setting, *Journal of Legal Medicine*, 17, 82–91.

Gosch, A. and Courts, C., 2019, On DNA transfer: The lack and difficulty of systematic research and how to do it better, *Forensic Science International: Genetics*, 40, 24–36.

Hanlon, M., 2011, As Amanda Knox walks free, now DNA evidence is on trial, *Daily Mail Online*, October 5.

Hansson, O., Finnebraaten, M., Heitmann, I.K., et al., 2009, Trace DNA collection – Performance of minitape and three different swabs, *Forensic Science International: Genetics Supplement*, 12, 189–190.

Hellmann, P., 2011, *The Helmann-Zanetti report*, On the acquittal of Amanda Knox and Raffaele Sollecito, Translated into English, December 16, 2011, available at http://www.hellmannreport.wordpress.com

Helmus, J., Bajanowski, T., and Poetsch, M., 2016, DNA transfer – A never-ending story, A study on scenarios involving a second person as carrier, *International Journal Legal Medicine*, 130, 121.

Helmus, J., Pfeifer, M., Feiner, L., et al., 2018, Unintentional effects of cleaning a crime scene – When the sponge becomes an accomplice in DNA transfer, *International Journal Legal Medicine*, 133, 759–765.

Helmus, J., Poetsch, J., Pfeifer, M., et al., 2020, Cleaning a crime scene 2.0- what to do with the bloody knife after the crime, *International Journal of Legal Medicine*, 134, 171–175.

Himmelreich, C., 2009, Germany's phantom serial killer: A DNA blunder, *Time Magazine*, March 27.

James, E. and Brown, N., 2022, High court rebuffs Neill-Fraser again, *The Canberra Times*, August 12.

Johannessen, H., Gill, P., Roseth, A., et al., 2021, Determination of shedder status: A comparison of two methods involving cell counting in fingerprints and the DNA analysis of handled tubes, *Forensic Science International: Genetics*, 53, 102541.

Lehmann, V., Mitchell, R., Ballantyne, K., et al., 2013, Following the transfer of DNA: How far can it go? *Forensic Science International: Genetics Supplement Series*, 4, e53–e54.

Mapes, A., Kloosterman, A., van Marion, V., 2016, Knowledge on DNA success rates to optimize the DNA analysis process: From crime scene to laboratory, *Journal Forensic Sciences*, 61, 4, 1055–1061.

Marasca-Bruno, 2015, *The supreme court of cassation motivation report*, available at http://www.amandaknoxcase.com/wp-content/uploads/2015/09/Marasca-Bruno-Motivations-Report.pdf

McGregor, K., 2015, Charges dropped against Adrian Drummond, who was jailed over alleged schoolgirl kidnap, *The Advertiser*, September 7.

Meakin, G. and Jamieson, A., 2013, DNA transfer: Review and implications for casework, *Forensic Science International: Genetics*, 7, 434–443.

Meakin, G., Butcher, E., van Oorschot, R., et al., 2015, The deposition and persistence of indirectly-transferred DNA on regularly used knives, *Forensic Science International: Genetics Supplement Series*, 5, e498–e500.

Meaney, T. and Schafer, S., 2016, The Neo-Nazi murder trial revealing Germany's darkest secrets, *The Guardian*, December 15.

Neill-Fraser v Tasmania {2019} TASSC 10.

Neill-Fraser v Tasmania {2021} TASCCA 12.

Otten, L., Banken, S., Schurenkamp, M., et al., 2029, *Forensic Science International: Genetics*, 43, 102126.

Poy, A. and van Oorschot, R., 2006a, Beware; gloves and equipment used during the examination of exhibits are potential vectors for transfer of DNA-containing material, *International Congress Series*, 1288, 556–558.

Poy A. and van Oorschot, R., 2006b, Trace DNA presence, origin, and transfer within a forensic biology laboratory and its potential effect on casework, *Journal Forensic Identification*, 56, 558–576.

R v Drummond [2013] Supreme Court of South Australia, SASCFC 135.

R v Drummond (No 2) [2015] Supreme Court of South Australia, SASCFC 82.

Spiegel online, 2009, Phantom killer is a phantom, March 27, available online at http://www.speigel.de

Spiegel online, 2012, DNA tests solidify suspicions in police killing case, August 13, available online at http://www.speigel.de

Steensma, K., Ansell, R., Clarisse, L., et al., 2017, An inter-laboratory comparison study on transfer, persistence and recovery of DNA from cable ties, *Forensic Science International: Genetics*, 31, 95–104.

Szkuta, B., Ballantyne, K., and van Oorschot, R., 2017, Transfer and persistence of DNA on the hands and the influence of activities performed, *Forensic Science International: Genetics*, 28, 10–20.

Szkuta, B., Ansell, R., Boiso, L., et al., 2019, Assessment of the transfer, persistence, prevalence and recovery of DNA traces from clothing: An inter-laboratory study on worn upper garments, *Forensic Science International: Genetics*, 42, 56–68.

Taupin, J. and Cwiklik, C., 2010, *Scientific protocols for the forensic examination of clothing*, CRC Press, Boca Raton, Florida.

Taylor, D., 2016, The evaluation of exclusionary DNA results: A discussion of issues in R v Drummond, *Law, Probability and Risk*, 15, 175–197.

Taylor, M., 2020, FBI study: M-Vac system collects more DNA than swabbing, *Forensic News*, August 3.

Temko, N., 2008, Germany's hunt for the murderer known as the 'woman without a face', *The Guardian*, November 9.

Templeton, J., Taylor, D., Handt, O., et al., 2015, DNA profiles from fingermarks: A mock case study, *Forensic Science International: Genetics Supplement Series*, 5, e154–e155.

Thompson, W. and Scurich, N., 2018, When does absence of evidence constitute evidence of absence? *Forensic Science International*, 291, e18–e19.

van Oorschot, R. and Jones, M., 1997, DNA fingerprints from fingerprints, *Nature*, 387, 767.

van Oorschot, R., Treadwell, S., and Beaurepaire, J., 2005, Beware the possibility of fingerprinting techniques transferring DNA, *Journal of Forensic Sciences*, 50, 6, 1–5.

van Oorschot, R., Ballantyne, K., and Mitchell, J., 2010, Forensic trace DNA: A review, *Investigative Genetics*, 1, 14.

van Oorschot, R., McColl, D., and Alderton, J., 2015, Activities between activities of focus- relevant when assessing DNA transfer probabilities, *Forensic Science International: Genetics Supplement Series*, 5, e75–e77.

van Oorschot, R., Szkuta, B., Meakin, G., et al., 2019, DNA transfer in forensic science: A review, *Forensic Science International: Genetics*, 38, 140–166.

Vecchiotti, C. and Conti, S., 2011, *The Conti-Vecchiotti report*, available at http://knoxdnareport.wordpress.com

Verdon, T., Mitchell R., and van Oorschot, R., 2014, Evaluation of tapelifting as a collection method for touch DNA, *Forensic Science International: Genetics*, 8, 179–186.

Chapter 4

Trace DNA Sources

BOX 4

- Skin contact.
- Non-self DNA on hands.
- Non-self DNA on body.
- Shedder status.
- 'Wearer' DNA on clothing.
- Background and prevalent DNA.
- Persistence of DNA.
- Recovery of DNA.

4.0 Introduction

Direct transfer of any DNA-containing material from a person may occur via various means such as spurting blood, spitting or coughing on an item or having contact with an item – such as smearing blood on clothing, or depositing DNA through wear from skin surfaces or handling the clothing or object. The terms direct and primary transfer have been used interchangeably in the literature and in common usage.

DOI: 10.4324/9781003158844-4

Secondary transfer has been used to imply indirect transfer in general but this is only partially correct. Indirect transfer encompasses not only secondary transfer but also higher levels such as tertiary, quaternary, quinary and so on.

The notation 'trace' implies that there is no visible deposit and/or the biological matter has not been sourced to a particular fluid such as blood – it is denoted only as 'DNA'.

A DNA deposit obtained from a surface may incorporate direct and indirect transfer events of DNA. Transfer from a hand contact, for example, may result in deposits of DNA of the person making the contact, as well as any other DNA that may have been present on the hand – deposited through a variety of ways that also may be direct or indirect.

Context is important as emphasized in this text.

Transfer can be *bi-directional.* If two surfaces come into contact there can be an exchange of DNA-containing material, following Locard's Exchange Principle. (See Figure 4.1.)

Transfer of DNA from a glove to an exhibit, if that glove had previously touched another item, is tertiary transfer and may not be realized as such because *the glove is not sampled for DNA*. Consequently investigator-mediated transfer of trace DNA may not necessarily reveal mixtures involving the investigator if that investigator wears suitable protective clothing. (See Figure 4.2.)

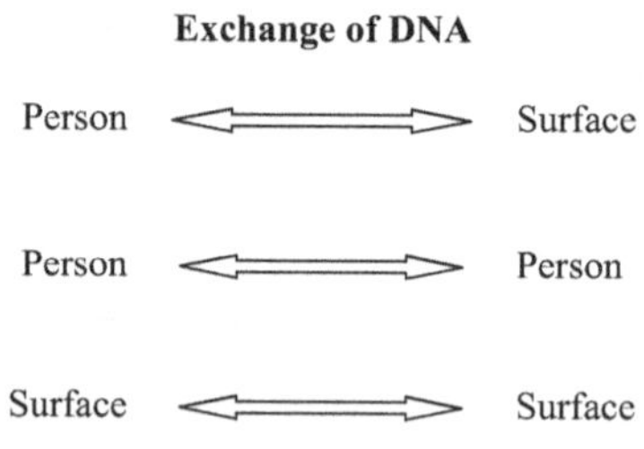

FIGURE 4.1 Bi-directional transfer of DNA.

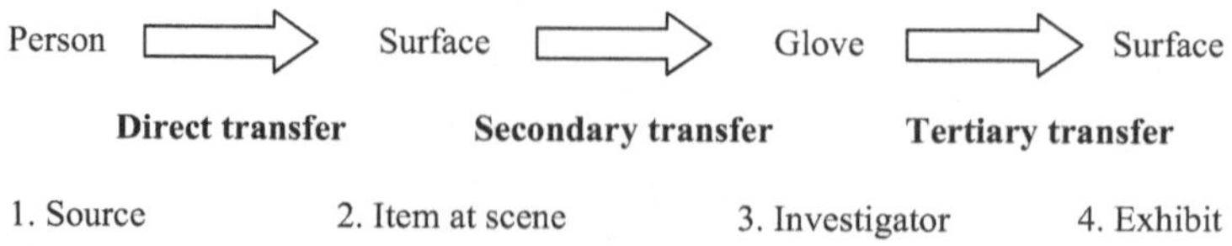

FIGURE 4.2 Investigator-mediated transfer of DNA.

4.1 Skin Contact

4.1.1 Human Skin

DNA from an individual – their own DNA or 'self-DNA' – through skin contact was initially believed to derive from epithelial cells shed or sloughed from the outermost layer of the surface of the skin. This was an explanation that accorded with the transfer of nucleated cells as expected with other biological matter such as blood. As of writing there is debate on the nature of the DNA that is in 'touch DNA' – DNA that is accepted to be from the self and from the skin. This text will refer to deposits designated as such in the literature with the proviso that this author disagrees with the term and should be denoted as 'trace DNA'.

Human skin is called the epidermis with multiple cell types and layers that protect the body against the environment. The outer layer of human skin is called the stratum corneum and is constantly being shed. This layer is keratinized and consists of corneocytes which are flattened cells without a nucleus full of protein called keratin. The innermost layer of the skin manufactures cells that migrate to the surface so that there is constant skin renewal.

Biological material transferred from the skin onto an object is now believed to comprise the corneocytes without a nucleus, undegraded or partially degraded epithelial cells as well as extracellular DNA (eDNA). The prevalence of each of these components, their biological and physical relationships within trace samples and their impact on total DNA yield is still unknown (Miller et al., 2021).

'Trace DNA' through skin contact is now accepted to contain a mix of shed epithelial cells and cell-free DNA with secretions such as sweat commonly used as the transfer vector. Trace DNA may also arise from bodily fluids or shed cells from different parts of the human body, depending on contacts made by the skin surface prior to the deposition.

4.1.2 Self-DNA from the Hands

There was an assumption that the propensity to leave behind genetic material from skin surfaces such as the hand reflected the shedding of the outermost layer of skin cells, and that assumption continues today in some arenas despite limited evidence to support it (Burrill et al., 2019). Recently, this assumption of the source of skin-derived 'touch DNA' has started to be questioned.

Data regarding DNA deposits allegedly from handled items and documented in the literature and in case reports had focused on DNA typing results, sometimes including the DNA amount quantified, but the microscopy or cellular characteristics of the deposit was not generally queried (ibid.).

An early study examining 'fingerprints' found that the majority of deposited cells were nuclei-free corneocytes with only a limited number of nucleated cells observed (Allessandri et al., 2003).

Later studies thought that extracellular body secretions such as sebum and sweat acted as vectors for the transfer of cell-free DNA through the layers of the skin (Quinones and Daniel, 2012; Zoppis et al., 2014).

Extracellular DNA was noted as a substantial proportion of total DNA recovered from handled substrates (Burrill et al., 2021).

A study (Miller et al., 2021) examined deposits from hands of 24 different participants in a trial where their hands gripped tubes for 5 minutes. Extracellular DNA yields from these deposits varied widely across individuals and from the same individual – for example, one individual had zero DNA quantified on one day to more than 9 nanograms (ng) on another. This may mean that the quantity of extracellular DNA deposited by touching from the skin may depend on extrinsic factors, and not a fixed 'shedder status' (see below).

Hands act as vectors for all types of material. DNA on hands may not only be from DNA material from the hands but other body fluids from the person including saliva, blood and semen. This is indirectly transferred self-DNA.

The spread of disease via hands is an illustration. A study on respiratory tract infections described that a substantial portion of human respiratory tract infection is thought to be transmitted via contaminated hand contact with the mouth, eyes and/or nostrils (Nicas and Best, 2008). This paper studied office workers in isolation from others and found that the average contact by hand to the eyes, nostrils or lips was approximately 15 times per hour. Another study determined that people performing everyday activities such as sitting at a café, on public transport, cooking, being at home or in a park touched items (and not specifically faces) 15 times per minute with their dominant hand (van Oorschot et al., 2015).

A further study (Fonnelop et al., 2017) supported this idea by noting that on average people performed approximately 15 touches every 60 seconds. The paper also noted that 'it is possible that people in close relation and/or sharing accommodation are more prone to transferring DNA from each other, even some time after their last contact'. Thus the prevalence of DNA of multiple people in office or home environments and public places is to be expected.

A combination of sources and factors influence DNA derived through touching with the skin (van Oorschot et al., 2019). Secretions from sebaceous glands in the skin which are abundant in the scalp, face and around the apertures of the ear, nose, mouth and anus, may act as a vector for DNA. Secretions are under hormonal control. Sweat glands are located on all areas of the body with the highest density in palms and soles of the feet. The overall quantity of DNA retrieved within samples taken from sweaty hands increased when cell-free nucleic acids could be detected in sweat (Quinones and Daniel, 2012).

An early study published a list of amounts of DNA recovered from bare hands or surfaces touched once with bare hands (Meakin and Jamieson, 2013). This amount varied widely, from 0 to 150 nanograms (ng). One cell contains 6.5 picograms (pg) of nuclear DNA; a picogram is one thousandth of one nanogram. Regular use of an item may also leave variable quantities of DNA. The review states that it is thus not possible, to establish from the amount of DNA recovered from a surface, whether the DNA was deposited by a singular touch or by regular use (ibid.).

It has been observed that less DNA was deposited by, or collected from, the entire palm area than by parts of the fingers themselves, despite the fact that palms have a larger surface area. Different parts of a hand, especially fingertips, appear to have proportionately more DNA, more non-self DNA and may be more likely to pick up and transfer non-self DNA than others (McColl et al., 2017). This study found a range of DNA quantities on the hands, varying from 0 to 585 ng.

A later study on fewer participants found a range of 0 to 23 ng in total quantity of DNA on hands (Samie et al., 2020).

Hand activities during robberies were analyzed by viewing CCTV video footage of both armed and unarmed robberies and attempted robberies (Stella et al., 2017). Approximately 50 offences inside stores were studied predominantly in the United Kingdom and the United States between the years 2006 and 2016. The number of touches with the dominant hand (around 5 per minute for armed robberies and 9 per minute during unarmed robberies) were lower than those during general activities (accepted to be 15 touches per minute). This emphasizes the complexity involved when trying to relate laboratory data to real-life situations.

4.1.3 'Self-DNA' from Skin from the Body

There have only been a few studies on the normal level of DNA found on the skin on areas of the body other than hands. The levels of DNA on the skin of children aged 0–5 years and the neck area of adults (Graham and Rutty, 2008) have been described, but these studies were performed using less-sensitive techniques than those used today. Further research has been recommended, especially on intimate areas of the body regarding sexual offence cases, for the prevalence of both self DNA and non-self DNA on these areas (van Oorshot et al., 2019). See Chapter 5 for discussion regarding intimate body samples.

4.2 Non-self DNA on Skin from the Body

4.2.1 Non-self DNA on Hands

The skin on hands or other skin on the body may contain 'non-self' DNA deposited through direct or indirect transfer from another individual. Studies often focus on 'hands' rather than other skin areas on the body because hands are most often used in 'activities' and activity-level analysis appears to focus on handling objects or touching with the hands.

Studies investigating the deposits of DNA from hands on objects through touching found that in addition to their 'self' DNA, foreign 'non-self' DNA from other persons was observed in the DNA profiles obtained. Non-self DNA can arrive on hands via direct transfer from other skin surfaces, or indirectly from contact with a surface on which DNA is present.

A reservoir of DNA from multiple deposits of the occupants may exist in home, social or work environments. This DNA can be picked up by hands, redistributed or added to by DNA on that hand. The transfer can be bi-directional.

Hands are an important vector or means of transfer of DNA. Hands and fingernails can act as ready vectors for the transmission of disease (in the medical context) or evidence (in the forensic context). Blood or other body fluid or cellular material may transfer from hand to hand in multiple transfer events, and hands and fingernails may also act as a 'reservoir' of DNA or body material. Hands and fingernails may transfer nasal secretions, saliva, fluid from the eyes/nose/mouth or body fluid from the wounds or orifices of the individuals themselves or from other individuals.

The persistence of body fluid or any DNA on hands or fingernails depends on the activities after the DNA or body fluid was deposited. It will also depend on the location and may remain relatively longer in an area where it is less likely to be dislodged, such as in crevices or underneath the nails.

A well-known case in the United States that graphically demonstrates how hand-to-hand transfer of DNA may occur via an intermediary item, and the ease of that transfer, is that of Lukis Anderson (Smith, 2016; Worth, 2018).

CASE 4.1 LUKIS ANDERSON

The only evidence against the accused in a murder case from the United States comprised DNA from underneath the fingernails of a deceased male in his own home, after a violent home invasion in California in 2012. DNA from a suspect Lukis Anderson, a homeless person, was proposed to match the foreign DNA found under the murdered man's fingernails. Months later after the imprisonment of the accused, it was found that Lukis had been unconscious in a hospital at the time of the murder, and five times over the legal alcohol limit for driving.

It was ultimately discovered that paramedics conveyed the unconscious Lukis to a hospital, from where he was found outside a liquor store. The same paramedics then attended to the murdered man at the crime scene. It was believed the same oxygen monitor probe had been placed on the fingers of both men. Thus the possible mechanism was indirect transfer of DNA from the accused through the probe placed on the finger of the accused by paramedics and then the finger of the deceased by the same paramedics.

Lukis Anderson was released from jail.

Note that in the above case the prior odds – see Chapter 2 – would be zero as Lukis Anderson was in hospital at the time of the offence. The posterior odds are also zero no matter what the likelihood ratio is of the DNA evidence.

One interesting study found 'non-self' DNA present in 97% of 64 samples of direct deposits by hands, after carrying out normal office activities, on DNA-free plastic knife handles. The hand donor was the clear major contributor in 83% of samples, but 12% of mixtures had no clear major or minor contributor (Samie et al., 2016).

DNA samples from 240 handprints were investigated from ten participants (Goray et al., 2016a). It was reported that the majority of the samples (79%) contained indirect transferred DNA, and in most cases this appeared as a minor contributor (≤ 30% of total DNA). However, the investigators also detected incidences of *the participant* appearing as the minority (≤ 50% of total DNA) or *excluded* from the obtained DNA profile in 7.5% of the samples, derived from designated 'poor or middle range' depositors (shedders).

A study examining handling and transfer of DNA to knives gripped by hands, found varying quantities of 'non-self' DNA on hands of the participants (Samie et al., 2020). The percentage of non-self DNA on the hands varied substantially between donors with an average of 8% non-self DNA; however for one participant the total quantity of DNA from the hand was only 70% of his own DNA (ibid.). Regarding the primary transfer of DNA on knife handles it was found that a total quantity of 0 to 5 ng was transferred from the hand to the knife handle. Again, however, there were differences for the donors as for one donor 40% of the total DNA transferred came from someone else (ibid.).

Another study examined prolonged handshaking (of 2 minutes) between person A to person B, when those two individuals had washed their hands and worn a glove 1.5 hours prior to shaking, thus maximizing the potential of DNA transfer from self DNA hand to hand (Cale et al., 2016a). Individual B then handled a knife and the study found detectable levels of person A were transferred to the knife in 85% of occasions and was observed as either the only or the major contribution in 25% of occasions (total of 25 subjects). A conclusion thus drawn was that a person may touch an object yet transfer none of their own DNA, but the DNA of others with which they have had social contact.

The finding of the above study provoked criticism in the literature (or feedback) due to the realism or otherwise of the scenario (e.g., Goray et al., 2016b; Cale et al., 2016b; Kokshoorn et al., 2016). However, the particular study provides valuable information on 'unknowns' and researching the limits of testing, as the study was purposely designed to maximize DNA transfer. Examiners need to be open-minded regarding the alternatives of indirect or secondary transfer. The authors of the study noted that at the time secondary transfer was not being discussed in the United States compared to the United Kingdom and elsewhere.

The criticism of the Cale et al. study showed the uncertainty with which examiners are faced and how they reason in the face of this uncertainty. This reasoning may, or may not, be translated to the criminal justice system or the court. An example of examination uncertainty in the criminal justice system was a High Court Appeal of Australia, appealed from a conviction in South Australia, and a subsequent Criminal Court of Appeal in the State (*R v. Sumner, R v. Fitzgerald*, 2013; *Fitzgerald v. The Queen*, 2014; Gans, 2014). The item in question was a 'didgeridoo', an indigenous Australian musical instrument.

CASE 4.2 TRANSFER OF DNA; TIME OF DEPOSIT; SOCIAL CONTACT

A man died and another had brain injuries, from a violent altercation in the south of Australia around 6.00 am in June 2011. A group of at least 6 men (possibly up to a dozen) had forced their way into a home and attacked two of the occupants, using weapons including axes and gardening forks. The trial before a judge and jury convicted Grant Sumner and Daniel Fitzgerald of murder and both were sentenced to life imprisonment.

There was no direct evidence that either man inflicted the fatal injury or injuries; no one saw the injuries occur and prosecutors could only name two of the multiple men who may have visited the premises. The prosecution of Daniel Fitzgerald relied on DNA obtained from a didgeridoo – there was no eyewitness evidence or any other evidence (thus 'sole-plank'). It was proposed that the DNA on the didgeridoo derived from Fitzgerald's blood and was transferred at the time of the attack. The defence argued that alternative hypotheses regarding the deposition of DNA were open. These hypotheses included a handshake between Fitzgerald and Sumner the previous evening at a boxing match. It was accepted that Sumner had visited the house during the night approximately 2 hours before the questioned altercation, had several physical altercations and received a fractured jaw.

On the night in question the man who was now deceased had played the didgeridoo. At some point it was then leaning on a freezer in the kitchen and it was briefly picked up by one of the female occupants of the house to defend herself during the attack, but she did not hit anyone. The musical instrument finally ended up in the living room next to the deceased and it was stained with what appeared to be blood.

The didgeridoo was alleged to have several DNA deposits – DNA matching that of the deceased, the other victim and several unknown people. There was DNA obtained from two tiny reddish brown spots. These spots were not examined in situ for blood pattern appearance (blood stain pattern analysis). In fact, the reporting scientist had never examined the didgeridoo, only photographs. The samples had been scraped off by others and then submitted for biological examination. The prosecution stated that it was open for the jury to conclude the method of deposition of the spots was airborne due to their very small size – as opposed to transfer or contact. Following a database search, the major component within the DNA profile from the deposit linked to Fitzgerald with a likelihood ratio of 200 million; an unknown source contributed to the minor component (Szkuta et al., 2018).

There were large areas of reddish brown staining on the didgeridoo with DNA that matched the DNA of the deceased and the other victim. The prosecution submitted that this inferred there was blood from the two tiny spots. The stains on the didgeridoo alleged to be blood were not confirmed as blood – the test was positive for a blood screening test only. It was said by the reporting laboratory forensic scientist that blood is a rich source of DNA. By way of contrast, trace DNA contact through skin cells is a very poor source of DNA.

The court stated that if the red spots were not blood, it was necessary to consider whether the DNA came from direct transfer or secondary transfer. The hypothesis of a handshake required secondary transfer. The laboratory scientist stated primary transfer was a much more likely source of DNA than that of secondary transfer. DNA matching that of Fitzgerald provided the major component of the DNA mixture profile from the 'spots'. The DNA matching that of Sumner was not in the stain or on the didgeridoo at all (*this author note: that was tested*). Secondary transfer was thus unlikely as the direct depositor of the DNA did not leave any of their own DNA behind and secondary transfer occurs infrequently. Note the scientist is commenting on likelihoods of the scenarios and stating explanations. However, the scientist also stated that DNA could not be 'dated', and they accepted that contact or trace DNA may have been on the didgeridoo some time before the attack. Indeed, they also agreed that the DNA may have been 'under the stain' – placed on the didgeridoo at an earlier time – and that the 'reddy-brown stains' may have contributed nothing to the DNA analyzed.

The Appeal Court in South Australia dismissed the defence transfer scenario of a handshake with the convicted man the previous night. The court decided that it was a 'succession of unlikely events' for the DNA of Fitzgerald to be transferred by secondary transfer. The conviction was reaffirmed. This verdict was then appealed to the High Court of Australia.

The appellant asked the High Court:

- Is DNA evidence alone sufficient to establish both presence and participation for the purposes of joint enterprise liability beyond a reasonable doubt in circumstances where the issue is not whether there is a match but when and how it got there?
- Was it unreasonable to convict the appellant in circumstances where an expert called by the prosecution to give evidence about DNA testified that: (1) the science concerning 'secondary transfer' of DNA was in its infancy; (2) there were no statistical studies about the frequency

of its occurrence; (3) in her reading of scientific literature, some put it at 1%, others 5% and others higher; (4) 'secondary transfer' might occur through a handshake; (5) it was not possible to date the DNA found; and (6) secondary transfer of DNA was possible in this case such that there was a reasonable hypothesis consistent with innocence?

The High Court was urged by the respondent to reject the secondary transfer theory, whether applied to Sumner or another, essentially on the basis that the appellant's DNA was the major contributor to the DNA in the questioned sample, and the likelihood that the appellant's DNA derived from blood. It was also submitted that whether the DNA in the sample derived from blood could be assessed against the 'unlikelihood' of a secondary transfer. The High Court noted that the degree of circularity in those submissions reflected the dearth of evidence of what had been done with the didgeridoo before the attack.

The High Court unanimously held that the prosecution's main contention, that the appellant's DNA in the sample obtained from the didgeridoo derived from his blood, was not made out beyond reasonable doubt. Further, the recovery of the appellant's DNA from the didgeridoo did not raise any inference about the time when, or circumstances in which, the DNA was deposited there. The court held that it could not be accepted that the evidence relied on by the prosecution was sufficient to establish beyond reasonable doubt that the appellant was present at, and participated in, the attack. The jury, acting reasonably, should have entertained a reasonable doubt as to the appellant's guilt. As the evidence was not capable of supporting the appellant's conviction for either offence, no question of an order for a new trial arose.

Case 4.2 raises many issues. The following are the findings in the scientific literature:

- Forensic examination of a stain or deposit requires the scientific method and includes evaluation of morphology, location and extent; a sampling decision can then be scientifically informed.
- Biological confirmatory tests are required to denote blood; all biological tests consume sample, including presumptive (which are really screening) tests.

- A DNA profile cannot inform us as to how it was transferred.
- A DNA profile cannot inform us as to its time of deposit.
- The absence of a person's DNA in a mixed profile cannot imply that person did not transfer the DNA.
- Commenting on likelihoods of direct transfer versus secondary transfer is commenting on the propositions, which is not the remit of the scientist.
- Speculation concerning whether the DNA came from blood or through direct or secondary transfer is not evaluation suitable for a court.
- If the deposit or stain is not initially scientifically evaluated then the context may be lost.
- A lack of initial evaluation may impact on the resulting quality of the DNA profile/evidence and interpretation.

This case has been used as an example for activity-level propositions and Bayesian networks with many assumptions (Szkuta et al., 2018). This Bayesian analysis found that the DNA evidence supported Proposition 1 that Fitzgerald was present in the house during the attack and held the didgeridoo, versus Proposition 2 that Fitzgerald was not in the house during the attack and someone else held the didgeridoo, with likelihood ratios varying according to the time interval considered – such as the findings are approximately 57 times more probable if Fitzgerald was present at the house than if he was not (ibid.). Note the original prosecution theory proposed in the case trial was that Fitzgerald's blood was deposited via airborne spatter on the didgeridoo, and there was no evidence that any of the attackers held the didgeridoo during the altercation – but a female occupant of the premises was believed to have held the didgeridoo.

The alternative 'activity level' scenarios proposed in the study examining activity-level propositions (Szkuta et al., 2018) did not actually address the propositions by prosecution and defence in the case. The authors of the study stated that as the case was selected only as an exemplar of issues facing forensic biologists, the circumstances and variables were not exactly replicated but modelled within the laboratory to provide indicative data in similar scenarios (ibid.). The didgeridoo was substituted by an axe handle to provide a broader application to the data collected. The authors noted a lack of data relating to the persistence of transferred DNA on hands over time, so a number of handshaking experiments based on the time intervals presented in the case were performed (ibid.).

The extent of non-self DNA present on a person's hands upon contact with items is dependent on the actions of the hands prior to sampling or depositing. Contradictory results have been obtained from studies and few studies have focused on the origin of non-self DNA on personal objects (van Oorschot et al., 2019). Experimental variables such as the use of pre-cleaned objects, using the same participants in the study but in different combinations, and low numbers of samples limit relevance to casework. More samples need to be collected in realistic scenarios that may require extensive ethical considerations (ibid.).

An awareness of known associates of an owner/user of an item or space may assist in understanding contributions to 'background DNA' (see below). During casework investigations comparisons to reference/elimination samples from known potential contributors may assist in the interpretation of mixture DNA profiles (van Oorschot et al., 2019).

This author has reviewed cases where the reference DNA profiles were initially not provided but requested by the defence, and subsequently agreed during court proceedings as potential donors of the deposit due to co-habitation. See Chapter 6 for further discussion.

A reservoir of DNA from multiple deposits of the occupants may exist in home, social or work environments. This can be picked up by hands, redistributed or added to by DNA on that hand. The transfer can be bi-directional.

4.2.2 Non-self DNA from Areas of the Body

There is comparatively little investigation into non-self DNA from areas of the body other than the hands.

Skin-to-skin contact, such as during physical assaults, has been investigated in the context of manual strangulation. Detectable levels of 'non-self DNA' are normally present on the surfaces of necks, especially when persons are co-habiting with other individuals. A study using less-sensitive techniques than today (Graham and Rutty,

2008) found non-self DNA on the necks of 14 of 24 volunteers, mostly from domestic co-habitants. Another study found 'non-self' DNA in 83% of 20 samples taken from the neck and 88% of 40 samples taken from the hands, originating from up to 3 or 4 contributors, respectively (van den Berge et al., 2016). It has been stated that any DNA obtained from a neck in a strangulation case may, or may not, be from the assailant (Meakin and Jamieson, 2013).

A more recent study examined non-self DNA on the neck during a 24-hour time period (Fantinato et al., 2022). It was accepted as a premise that non-self DNA is present on the skin due to DNA transfer during daily activities, and that high quantities could be present on the neck from co-habiting individuals. Often during criminal investigations the accused and complainant co-habit or are in a relationship and thus DNA could be innocently transferred at a time unrelated to the crime event (ibid.). The study investigated levels of non-self DNA on the neck over a 24-hour time period since washing, from 120 samples gathered from 20 adult volunteers. DNA concentrations ranged from 0.0004 ng/ul to 4.69ng/ul, a difference of more than 10,000 fold.

Full and partial donor profiles were obtained from 38% of the samples, while 53% of the samples resulted in 2 person mixtures and 9% in 3 person mixtures. All 2 person mixtures were donor/partner except for 1 sample, where the displayed alleles came from an unknown contributor. The 3 person mixtures comprised donor, partner and an unknown contributor. The sample donor was the major component in all but 3 DNA mixtures. The DNA quantity of the samples collected from the neck showed both inter- and intra-personal variations, with 1 sample generating no DNA profile. All of the 11 samples that displayed unknown DNA were from participants who used public transport, attended a public space, had social contacts or used unwashed clothing or items before collecting the samples. The conclusion of the study was that non-self DNA is normally present on the neck surface of adult individuals, due to DNA transfer during daily activities (ibid.).

4.3 Shedder Status

Shedder status, such as a person being a 'good shedder' or 'poor shedder', has been a term used to categorize a person with respect to the DNA deposited on a surface when touched with the skin, usually associated with hands. 'Shedder status' is the propensity for an individual to leave behind DNA material on surfaces. Many studies have purported to show that some individuals shed more DNA than others (Jansson et al., 2022).

A couple of decades ago it was believed that some individuals may have an increased propensity to shed DNA-containing cells compared to the rest of the population; these people were called 'shedders' (Lowe et al., 2002). Knowledge of an individual's shedding characteristics was thought useful in providing general background data in the interpretation of DNA trace evidence (Phipps and Petricevic, 2007). Other authors (Buckingham et al., 2016) concluded that a variety of factors influence the amount of DNA deposited by an individual, including shedder status.

Consistency was observed between the quantities of DNA deposited over time from individuals considered to be good shedders and those who were considered poor, while a good shedder was more likely to be identified following a transfer event compared to a poor shedder (Szkuta et al., 2018).

Studies have been performed to investigate factors that could influence a person's 'shedder status' such as skin diseases, presence of sweat (cell-free DNA in sweat), DNA in sebaceous fluid and thickness of the skin area which comes into contact with an object. In addition, the source of the DNA profile could be other cell sources that have been loaded onto hands during daily activities (this is non-self DNA). Further, personal habits such as touching the face and frequency of hand washing may also influence a person's shedder status (Fonnelop et al., 2017). Skin diseases that increase skin cell turnover may result in a higher deposition of DNA, for example patients with atopic dermatitis or psoriasis (Kamphausen et al., 2012), as the following case describes (Barnes, 2012; Gill, 2019).

CASE 4.3 DAVID BUTLER

Local English taxi driver David Butler's DNA was allegedly recovered from the fingernails of murdered sex worker Anne Marie Foy. It was presumed that Ms Foy had torn at his skin as he hit and strangled her, before dumping her body in a park near

Liverpool city centre in September 2005. The amount of DNA found by police was supposedly 'tiny', but enough to generate a hit against the United Kingdom's DNA database, identifying Mr. Butler as the source. His DNA was on the database due to willingly giving it to help solve a break and enter at his mother's house when he was living there. He denied ever having met the victim, but even though other evidence was lacking, the DNA evidence was enough to see him charged with murder.

Mr. Butler's defence team queried precisely how that DNA got onto the victim's nails. They established that he was sometimes known by the nickname 'flaky', because of the dry skin condition he suffered from, and suggested that perhaps some of his skin cells had transferred to bank notes that were later used to pay Ms Foy – an example of *secondary DNA transfer – or through other innocent means*. The victim also had glitter fingernail polish which is attractive to dust and DNA. Mr. Butler was acquitted at trial after 8 months on remand.

There are currently no standard quantities of DNA left on a specific area when touched in a standardized manner to help categorize an individual as a poor, intermediate or good 'shedder' (van Oorschot et al., 2019).

Defining high and low shedders is conditioned on having a reference group subject to equal sampling and processing conditions. Samples collected and analyzed under different conditions may not be representative. More research has been stated to be required to refine a reproducible shedder test, but it has still been suggested that this information will be useful for the evaluation of evidence at the 'activity level' (Fonnelop et al., 2017). This paper found that the shedder rate of an individual influenced the transfer rate of DNA to a T-shirt in a simulated attack scenario. There was a significantly higher amount of DNA deposited by high shedders compared to low shedders, and indirectly transferred DNA was more likely to be detected on the low shedder's T-shirt compared to the T-shirt of the high shedder (ibid.).

It is not completely understood why individuals have different tendencies to deposit DNA. The presence of transferred nucleated cells or cell-free DNA on the hands may occur because of habits: for instance touching the face or forehead could affect a person's shedder status (Johannssen et al., 2021). One study proposed that the amount of self-DNA deposited from hands is highly influenced by the individual levels of accumulated facial DNA, and that cells/DNA are often transferred to hands by touching or rubbing the face (Jansson et al., 2022).

Two different methods in evaluating shedder status – fluorescent cell counting and hand holding a tube – did not necessarily result in the same classification for each individual (Johannessen et al., 2021). Another study (Manoli et al., 2016) found that 77% of their participants changed shedder status if replicates were considered separately. This emphasizes the challenge of determining a person's shedder status and the importance of carrying out several trials to take into account a person's intra-individual tendency in DNA deposition.

More recent studies however have shown that individuals who deposit large amounts of DNA by touch one day could deposit undetectable amounts of DNA the next (Miller et al., 2021; Samie et al., 2020).

Such varying values of deposition of self DNA (cellular and extracellular) calls into question the notion of fixed 'good' and 'bad' shedders, at least for DNA deposited via the hand.

Instead, it has been recommended that shedding ability is characterized by a distribution (Samie et al., 2020). This study also stated that 'transfer probabilities' (this is for activity-level evaluation) may depend on the participant and also depend on the type of transfer (direct versus indirect). This means it is not possible to resort to a quantification of DNA on the one hand and assess what will be transferred to a surface. The distribution will be dependent on the donor, the substrate and the transfer mechanisms (ibid.).

4.4 'Wearer' or 'User' DNA on Clothing or Other Personal Items

Personal items and clothing are often collected as evidentiary or reference material, and often assumed to provide the DNA profile of the regular wearer or user. Clothing from a person wearing it at the time of the alleged offence may provide not only reference profiles from the wearer but also deposits occurring during the alleged offence. Personal items such as toothbrushes or hairbrushes may be used for potential reference profiles in missing person cases, in context.

4.4.1 'Wearer' DNA on Clothing

'Wearer' DNA is that DNA considered deposited during the everyday wear of an item of clothing, shoes or even bedding (Taupin and Cwiklik, 2010). Now it is considered as a term that is accepted by all parties as DNA from clothing worn by a person that may be used to 'condition' mixture DNA profiles, if there are indeed DNA mixture profiles on the garment. If the assumed wearer of the garment has all their DNA components present in the mixture DNA profile from the garment, then this may be a useful assumption for conditioning calculations.

Analysis of potential 'wearer' DNA on clothing may be attempted in cases where the body is absent, such as missing persons, or the body is too decomposed. Sometimes this is called 'background DNA' on the garment (see next section). It is important to note that garments may be shared, or used for other purposes than wearing.

It is usually believed that DNA is obtained from a garment since the last wash, but one study found that underwear worn by female children and washed with the rest of the family laundry had 'background' DNA on freshly washed clothes from the family (Noel et al., 2016). See Chapter 6 for further discussions on washing clothes.

While clothing and bedding can be analyzed for potential DNA from the 'wearer' or sleeper, it is not possible to use the reverse logic and denote any DNA found on a garment is from the wearer. Context is important.

The deposition of 'wearer' DNA is variable and depends on many factors such as duration of wear, the manner of contact with skin or body fluids and the construction and composition of the clothing or bedding. There is currently insufficient empirical data regarding textiles and deposits of trace DNA to help understand the various factors that may influence its transfer and detection. Understanding the construction and composition of clothing and other textiles is necessary to evaluate the various factors that may impact the detection of DNA and other deposits on clothing (Taupin and Cwiklik, 2010).

A level of DNA may exist on clothing from social and domestic inhabitants of an environment.

Deposition of 'wearer' DNA may or may not be visible. Soiling or staining on the garment may indicate potential areas to sample, since sweat and sebum are vectors for the transfer of DNA from the skin. Sweat and sebum deposits may produce a yellowish or whitish stain on garments depending on the garment composition.

It is accepted that the internal areas of garments such as inside collars and cuffs are targeted for 'wearer' DNA, as these areas have the greatest contact with skin. A study on worn garments found that the wearer of the garment was found in all of the detectable DNA profiles from the inner collar, with the major or sole DNA contributor identified (Magee et al., 2018). The amount of wearer DNA recovered from collars ranged widely from 1 to 83 nanograms (ng), and the amount of non-wearer DNA recovered ranged from 0 to 4.7 ng. The study described this non-wearer DNA as 'background DNA' as the garments in the study only had one wearer. The wearer was detected in all samples from the inside cuffs with interpretable DNA results (two samples had no DNA detected at all despite being worn for 2 hours) but most were mixtures with the major profile corresponding to the wearer. However, for one sample the amount of non-wearer DNA exceeded that of the wearer, with 57% coming from one non-wearer and a further 6% from a second non-wearer. Non-wearer DNA from the collars and cuffs was associated with family members of the wearer (ibid.).

DNA from close associates of the regular wearer may be found on clothing when sampling for 'wearer DNA' and this means that a garment does not have to be worn when detecting DNA. Family members, domestic co-inhabitants and the DNA of work colleagues have been found on clothing of volunteers during experimental studies. Studies on the prevalence and origin of 'background' DNA on worn upper garments found that 31% of 448 DNA profiles obtained from different areas on garments had contributions from family members and work colleagues (Szkuta et al., 2019). While the wearers of clothing always contributed DNA to the inside of collars, their close associates often contributed DNA to the outside surfaces of garments (Szkuta et al., 2020).

Spending time in a shared space such as a work office may result in the transfer of DNA, from others who had occupied that space, to garments. This transfer may also occur in other shared spaces such as motor vehicles (De Wolff et al., 2019). A study on 'touched' garments that had been worn by another individual showed that reportable background DNA (non-wearer and 'non-toucher' DNA) occurred in 14% of samples and may affect the assessment of who wore the garment (Breathnach et al., 2016). Consequently, the finding of DNA on clothing and bedding requires context and caution in interpretation and may not necessarily be from the 'wearer'.

Nucleated cells from other body areas such as the eyes, nose or mouth that may be transferred to clothing during wear may also yield successful DNA profiles. The hands may act as vectors of transmission of these cells to different parts of the clothing. Clothing that may yield successful DNA profiles due to regular contact with the skin, or mouth and nose secretions, include socks, gloves, baseball caps (sweat band), balaclavas, shirt underarms, shoe laces and the inside edge of the fly of underpants (Wickenheiser, 2002). Even lipstick has been found to transfer sufficient DNA from the 'wearer' of the lipstick.

4.5 Background DNA

4.5.1 Definitions

The DNA already present on a surface prior to the deposit of DNA of interest may be called 'background DNA'. Trace DNA cells are ubiquitous material that can be found for reasons unconnected to a crime event. It is thus reasonable to assume that most surfaces and items will have some DNA on them.

'Background DNA' has been defined as DNA that is not crime related but is present at a crime scene before the crime takes place (Fonnelop et al., 2017). It can originate from known and unknown individuals and can be propagated either by direct or by indirect transfer. It is important to recognize the potential or actual prevalence of DNA in the environment of the exhibit at a crime scene.

More recently there has been a distinction between 'background DNA' and 'prevalent DNA' (Gill et al., 2020). 'Background' DNA has been described as DNA that is present from unknown sources and unknown activities. 'Prevalent' DNA is that from known sources and activities where the analyst has a prior expectation of finding DNA from specific individuals, such as 'wearer DNA' from the garment of a known wearer, or DNA from a surface where there are known occupants of premises.

When the prevalence of DNA from a person of interest within a crime scene is relevant, it may also be relevant to collect samples such as a swab adjacent to the target area, similar items such as knives from a kitchen, and other items in the immediate vicinity – for example, a pillow under which a gun was found (van Oorschot et al., 2019).

In addition, it is important to obtain relevant samples of known users or wearers of an object that may assist the interpretation of any DNA profile. This is irrespective of them being the victim, the person of interest, or an incidental non-associated individual (ibid.). By acknowledging this, it also acknowledges that the DNA on an exhibit – purported to be instrumental to a crime – may have nothing to do with the crime.

4.5.2 Analysis of Background DNA

The persistence and preservation of DNA on environmental surfaces also depends on the quality and quantity. Trace DNA samples may suffer from the problem of low quantity and/or DNA that is degraded in some way, or not able to be extracted from the material on which it is deposited.

If the deposit of interest contains a good quality and large quantity of DNA, such as a fresh visible blood stain, then this DNA may overwhelm any 'background DNA' already present on the item. However, if the deposit of interest is trace DNA from a non-visible stain, the DNA from the deposit may be of a similar order, or in some cases is overwhelmed by background DNA.

The sensitivity of the more recent testing kits may add complexity in this way. Because the testing kits can analyse lower levels of DNA, extra caution is required.

4.5.3 Homes and Offices

Premises in which people occupy may have a reservoir of DNA from residents or visitors that could be transferred to evidentiary items. Further, when individuals have been in direct or indirect contact with each other the possibility of transfer of DNA should be considered from one individual to another and then to an item of interest, not involving a crime but part of everyday life. This 'background DNA' is expected to be present at any location with human traffic. Thus it may be necessary to take additional samples from crime scenes or items as 'controls' (see below).

One study examined levels of background DNA on flooring within houses (Reither et al., 2021). Swabs were taken directly from floors or from cotton cloths contacting flooring in various rooms of five different houses. DNA was detected from 97% of the samples directly contacting the floor and from 85% of cotton cloths, with interpretable DNA profiles from most. DNA is thus prevalent on flooring throughout houses and transferred readily from flooring to a contacting surface. The primary user of a space was usually, but not always, observed as a contributor, and occupants of houses were detected in DNA profiles collected from inside rooms where they had never been.

A Norwegian study (Fonnelop et al., 2017) noted that secondary transfer may occur if two relevant individuals have both occupied the same area before the crime event. A participant in the transfer study in an office was found to have DNA on the front of her clean T-shirt provided for the experiment, that corresponded to DNA from a person in the office who had been on leave for *2 months*, and who had not visited the office during that time. The modes of transfer could not be determined because the original deposit or background DNA was not located.

4.5.4 Motor Vehicles

The prevalence of DNA from drivers, passengers and other individuals on a wide range of locations within vehicles driven regularly by one individual has been explored in studies (e.g., De Wolff et al., 2021). DNA was able to be collected from all targeted sites, and yields and profile compositions varied between sites, with relatively higher yields from steering wheels and seats. The authors found that the driver was always observed as a contributor in DNA profiles from sites on the driver's side, in most instances being the sole or a major contributor, and the driver was also observed as a sole or a major contributor at several sites on the passenger side. DNA profiles of known recent passengers, close associates of the driver, and unknown individuals were observed on many of the sites on both the passenger and driver sides.

Surfaces in motor vehicles are thus expected to have quantities of DNA from regular occupants and their social or domestic contacts.

4.5.5 Controls

The 'hidden perpetrator effect' has been proposed (Gill, 2014). DNA will not always be recovered from the offender at a crime scene. If the perpetrator DNA is missing at the scene then donors of background and investigator-mediated contaminant trace DNA will automatically become suspects. An investigator may believe that a DNA profile recovered from a crime scene must have something to do with the crime. Consequently, the true perpetrator may be 'hidden'.

Every crime scene will have its own specific background, influencing transfer in that specific case, and therefore it is important to recognize the limitations of transfer probabilities. Knowledge about the natural background at a crime scene could be of value, but is not usually available (or indeed collected). However, it may be wise to collect such information from crime scenes in order to better inform 'case-relevant' probabilities in relation to activity-level propositions, and this could be an important topic for future research (Fonnelop et al., 2017).

These background samples could be considered as 'controls' for the exhibit from the environs from which the exhibit is collected. Taking a sample adjacent to the location where the exhibit was collected and analyzing for DNA may inform the interpretation of DNA on the exhibit.

The itinerant nature of DNA has been said to have serious implications for forensic science – if our DNA can make its way to a crime scene that we have never visited, are not we all suspects? (Worth, 2018).

4.6 Persistence of DNA

4.6.1 Introduction

It is not currently possible to determine the 'age' of a DNA profile from perusal of the DNA profile itself. That is, when the DNA was deposited on the surface from which the sample was collected. The time of deposit is intertwined with the variable of *persistence* in forensic DNA literature studies.

Little is known about the *persistence* of DNA on exhibits stored under different conditions (Szkuta et al., 2019). Exhibits may remain in police custody before forensic examination for considerable periods depending on examination backlogs, and can more than exceed a year at room temperature if the exhibit is considered dry (author experience).

DNA is a stable molecule and does not degrade in non-hostile conditions. This stability is one of the reasons for the replacement of the traditional blood grouping and enzyme testing originally employed in forensic laboratories. The rationale for the examination of a 'cold case' relies on the fact that DNA material persists over time. However, the amount that is detected is dependent on a variety of factors which are often unknown. It is always worthwhile to investigate further if it is thought the original police exhibits, such as clothing from the victim, have been destroyed. Medical swabs or debris collected from the items may still remain in case files or exhibit rooms or refrigerators. The stability of DNA, whether nuclear or mitochondrial, means there is potential for obtaining a result of evidential value many years later from the date of the original crime.

4.6.2 Persistence and Time of Deposit for DNA from Different Body Sources

The question of how long a DNA deposit may persist at a certain location is an important consideration in any forensic DNA analysis. This question may be difficult to answer as it depends on multiple variables such as the source of the DNA, the location and the conditions at the crime scene.

The persistence of spermatozoa and subsequently DNA in the internal human orifices such as the vagina has been a matter of interest for decades. Procedures to examine and collect evidence from sexual assault complainants have relied upon survival rates for spermatozoa in the internal cavity. The implementation of more sensitive DNA profiling technologies has changed perspectives and is discussed in Chapter 5.

A preliminary investigation of the persistence of trace DNA at crime scenes was published in 2009 (Raymond et al., 2009). This study investigated known amounts of biological material such as 'naked DNA' placed on external wooden painted windows of a residential building and collected after different times. DNA was negligible after 6 weeks – the longest period tested – on outdoor surfaces but full DNA profiles were obtained from items stored inside. Amounts of DNA recovered were examined at daily intervals. While there was a high degree of variability in the amount of DNA detected, an examination of the findings in their study show that very low (or even no detectable) levels of degradation occur within the first few days. The large number of factors affecting the recovery of DNA from casework samples greatly inhibits the accurate determination of the effect of a single variable such as time. These variables are difficult, if not impossible, to ascertain or control in case work.

Studies on the persistence of trace DNA have still been few until recently. The influence of weather on DNA stability has also been little investigated. A recent study from Singapore, and thus conducted in tropical conditions, showed that there was a high degree of variability of DNA persistence on items left outdoors and mostly depended on the amount of rain (Lee et al., 2019). The DNA from items placed indoors at ambient temperature was comparatively stable.

A study published in 2022 investigated outdoor and indoor scenarios for varying periods of time up to 12 months (Poetsch et al., 2022). Epithelial abrasions, blood cells and saliva cells were investigated in both indoor and outdoor scenarios under humid temperate climate conditions (in Germany). The study comprised samples from two individuals and the epithelial cellular deposits were obtained by rubbing cloth or plastic over the neck for approximately 5 seconds with medium pressure. Complete DNA profiles were obtained in 861 samples (58% of all samples). It was noted that *partial DNA profiles do not offer enough certainty to allow a bio-statistical calculation in Germany* (this author italics).

Indoor studies from the above paper showed that after up to a 9-month exposure, nearly all blood and saliva samples stored in the dark resulted in complete DNA profiles. However, after 12 months' exposure only 50% of blood samples and 75% of saliva samples demonstrated complete profiles. This was explained as possibly due to the use of plastic as supporting material, flaking and small amounts deposited. Epithelial abrasions showed a greater loss and after 3 months only half of these samples demonstrated a complete DNA profile, and none after 12 months even in the dark. It was postulated that the different composition of the bacterial fauna on the skin compared to, for example, bacterial fauna in saliva could have an impact (ibid.).

Influence of sunlight on indoor scenarios on blood and saliva samples resulted in complete profiles up to 9 months of storage. After 12 months of storage, only 13% of saliva samples demonstrated full profiles and 50% of all blood samples. The destructive effects of UV radiation on DNA are accepted and used for removal on implements but terrestrial UV light alone may only have a minimal effect (ibid.).

Outdoor scenarios showed a greater impact on obtaining full DNA profiles than indoor scenarios. A 'tipping point' of 3 months' exposure time for blood and saliva samples on cloth and soil was found (ibid.). Epithelial abrasions rendered no results as early as 2 weeks' exposure. After 12 months, no outdoor sample showed a complete DNA profile. Samples from summer scenarios demonstrated more complete profiles than winter scenarios, regardless of the source of DNA. The ultimate conclusion was that the influence of lower temperatures and higher humidity seems to outrank that of sunlight. Mild temperature and high humidity may favour microbial colonization (ibid.).

4.6.3 Persistence of DNA After Use

One article (van Oorschot et al., 2014) described the persistence of DNA deposited by the original user on objects, after subsequent use by a second person. The conclusion was that the degree of persistence of DNA depends on the type of object, the substrate it is made of, the area of the object targeted for sampling and the duration and manner of contact by a subsequent user. The original finding (described in Chapter 3) that the strongest DNA profile of samples taken from an object held by multiple individuals was not always from the last person who held the object (van Oorschot and Jones, 1997) is still true, despite further research and literature.

However, there is still a belief in some circles that 'DNA from the last handler, user, or wearer will be detected as the major contributor' (Meakin et al., 2021). Persistence of the original user and detection of subsequent users is dependent on a multitude of factors including the above, the ability to deposit DNA, and also the detection methods employed.

One paper (Fonnelop et al., 2015) studied the persistence and secondary transfer of DNA from previous users of equipment. The study showed that it was possible to transfer an initial user's DNA from computer equipment to the hands of a new user up to 8 days after receiving the equipment. Sampling between the keys of the keyboard identified the initial user's DNA to the end of the study, which was 42 days in length. Background samples from the keyboards before the start showed that in addition to the previous user, DNA from unknown persons was present. This shows the importance of sampling for 'background DNA' to assist in evaluating the persistence of trace DNA.

The results of another study (Meakin et al., 2017) showed that regular handling of knives could give detectable levels of regular user DNA that persisted for at least a week, varied between individuals and did not always result in complete DNA profiles.

A more recent study examined DNA on items by 'handling' (Atkinson et al., 2022). They found that the quantity of DNA deposited on 66 assorted items that are routinely submitted to the forensic laboratory for examination varied greatly for an experiment with a 'habitual user' handling the item for a week. One glove had just 0.012 ng quantified and one kitchen knife had 150 ng. Given that the threshold in the particular laboratory for amplification was 0.24 ng only 61 items proceeded to amplification. Note that laboratory methods may have different thresholds in different jurisdictions or between laboratories which may impact on relevance.

Amplified samples from the study (ibid.) contained DNA from solely the habitual user (13/61 results) which is direct transfer. The habitual user plus one unknown individual were found in 28 of 61 results and the habitual user and two unknown individuals in 21 of 61 results – examples of direct and indirect transfer in the one deposit.

Another experiment was performed by the same authors (ibid.) to investigate the DNA recovered following regular use by one handler over a week, and subsequent one-time use by another individual. Contributions were observed by the habitual user again in all amplified samples, along with up to 3 additional individuals – the one-time user (12/62 results); the one-time user and one unknown individual (25/62 results); one unknown individual (14/62 results); two unknown individuals (4/62 results); or a 4-person mix of habitual user, the one-timer plus two unknown individuals (4/62 results). The one-time user was the major donor in 10 samples but was not detected at all in 21 of the 62 samples.

Notably for the issue of indirect transfer, in the 47 samples where one or more unknown donors were observed, they accounted for DNA quantities ranging from 0.02 ng (one backpack) to 11 ng (one cell phone). Firearms had the highest average relative contribution of DNA from unknown individuals – 28% (ibid.). The study raised the important cautionary note that in most samples the examined *DNA was detected from individuals who did not participate in the known handling events* (this author's emphasis). The study also notes that although long known, it is important to ensure that 'investigators and prosecutors do not fall into the trap of assuming that the mere presence of an individual's DNA on an item provides strong support for the assertion that they directly contacted the item' (ibid.).

The study found tremendous variation regarding the amounts of DNA observed and the composition of DNA profiles. The study attempted to identify trends in differing shedder status or hygiene habits – but for the most part was unable to do so. In no instance was one individual consistently observed to deposit a higher or lower-than-average quantity of DNA on the items they handled. Further, in stark contrast to the habitual user, the odds are only slightly greater than even of detecting the one-time user's DNA following habitual use by another individual (ibid.).

Another finding was that there were many instances (34%) where the DNA of an individual known to have contacted the item was not observed as a component of the corresponding DNA profile, that was said to *confirm the oft-quoted adage* that the absence of an individual's DNA cannot be used to draw conclusions that the person did not handle or make contact with the object (ibid.).

4.6.4 Persistence of DNA after Cleaning/Washing

Initially discussions involving washing or cleaning of items seized for forensic examination focused on clothing. This was due to the belief by police that potential offenders may try to hide evidence on their clothing by cleaning or washing.

Studies focused on the persistence of transfer of body fluids, particularly semen and blood. Reportable DNA profiles obtained from semen stains on clothing were found after machine washing but also on co-washed clothing and indeed the washing machine drum (Noel et al., 2016). Similarly, DNA profiles were obtained from washed blood-stained cloths and from co-washed clean cloths (Edler et al., 2017; Kamphausen et al., 2015). These findings highlight the caution required when evaluating washed clothing. See Chapter 6 for further discussion on clothing items and washing, and potential for finding DNA from other sources than the clothing in question.

The removal of DNA by cleaning and the persistence of any remaining DNA despite a cleaning process on other items than clothing has been investigated. One study investigated whether DNA traces on different objects (knives, plates, glasses and plastic lids) can persist on the surface despite cleaning by hand washing or a dishwasher (Helmus et al., 2020). Samples taken after rinsing or hand washing resulted in complete DNA profiles 62.5% of the time, while cleaning in the dishwasher rendered everything almost completely DNA-free. Further, a carry-over of DNA traces from blood and saliva samples on plates was demonstrated up to a fifth hand-washed item.

A study investigated whether DNA traces could be distributed by cleaning an object (Helmus et al., 2019). A large table surface and fabric piece were artificially provided with saliva and blood in two series of experiments, and then wiped off with water or with soap water. There was a clear carry-over of DNA traces – all of the blood samples and three-quarters of the saliva samples led to a complete profile. The cleaning agent was of less influence unless it was a chloric agent which rendered everything completely DNA-free. Consequently, distribution of traces of DNA by wiping or scrubbing an object was demonstrated (ibid.).

4.7 Priority-based Recovery, Detection and Analysis

4.7.1 Principles

A DNA profile that may be attributed to a contributor/person cannot be recovered unless it has been transferred, persisted upon a surface and subsequently recovered in sufficient quantity so that it may be visualized in an electropherogram. The electropherogram is a graphical tool dependent on fluorescence that depicts sizes of fragments of DNA found in the nucleus of a cell, with these fragment sizes denoted from comparing references. This fluorescent depiction is dependent on other factors that precede it, such as copying, which may not be complete or result in artefacts, so consequently this is a 'picture' that may not be fully representative of what is actually there.

Searching and impact-based recovery techniques should be in a sequence from least destructive to most destructive. Ideally, the use of alternative light searching as well as visible light searching with a stereomicroscope should be employed first, as this technique is non-destructive. See Taupin and Cwilkik, 2010 for a discussion on examination, detection and recovery techniques for the examination of clothing which can be extrapolated to all forensic exhibits.

The use of a non-destructive screening/presumptive test (such as light sources) prior to any consumption of material using a presumptive or confirmatory test provides a useful indicator to the examining scientist. Where there are minimal or minute observable blood stains or semen stains, for example, and where the source-level link between a confirmed blood/semen sample and a DNA profile is imperative in case reporting, there should be a bypass in presumptive testing which may result in the consumption of part or all of the stain, and 'confirmatory' tests should be employed.

References

Alessandrini, F., Cecati, M., Pesaresi, M., et al., 2003, Fingerprints as evidence for a genetic profile: Morphological study on fingerprints and analysis of exogenous and individual factors affecting DNA typing, *Journal of Forensic Sciences*, 48, 3, 586–592.

Atkinson, K., Arsenault, H., Taylor, C., et al., 2022, Transfer and persistence of DNA on items routinely encountered in forensic casework following habitual and short-duration one time use, *Forensic Science International: Genetics*, 60, 1–14.

Barnes, H., 2012, DNA test jailed innocent man for murder, *BBC News*, August 31.

Breathnach, M., Williams, L., McKenna, L., et al., 2016, Probability of detection of DNA deposited by habitual wearer and/or the second individual who touched the garment, *Forensic Science International: Genetics*, 20, 53–60.

Buckingham, A., Harvey, M., and van Oorschot, R., 2016, The origin of unknown source DNA from touched objects, *Forensic Science International: Genetics*, 25, 26–33.

Burrill, J., Daniel, B., and Frascione, N., 2019, A review of trace 'touch DNA' deposits: Variability factors and an exploration of cellular composition, *Forensic Science International: Genetics*, 39, 8–18.

Burrill, J., Rammenou, E., Alawar, F., et al., 2021, Corneocyte lysis and fragmented DNA, considerations for the cellular component of forensic touch DNA, *Forensic Science International: Genetics*, 51, 102428.

Cale, C., Earll, M., Latham, K., et al., 2016a, Could secondary DNA transfer place someone falsely at the scene of a crime? *Journal of Forensic Sciences*, 61, 1, 196–203.

Cale, C., Latham, K., and Bush, G., 2016b, Authors' response, *Journal of Forensic Sciences*, 61, 5, 1399–1400.

De Wolff, T., Aarts, L., van den Berge, M. et. al., 2019, Prevalence of DNA in vehicles: linking clothing of a suspect to car occupancy, *Australian Journal of Forensic Sciences*, 51, S1, S103–S106.

De Wolff, T., Aarts, L., van den Berge, M., et al., 2021, Prevalence of DNA of regular occupants in vehicles, *Forensic Science International*, 320, 110713.

Edler, C., Gehl, A., Johlwagner, J., et al., 2017, Blood trace evidence on washed textiles: A systematic approach, *International Journal of Legal Medicine*, 131, 4, 1179–1189.

Fantinato, C., Gill, P., and Fonnelop, A., 2022, Non-self DNA on the neck: A 24 hours time-course study, *Forensic Science International: Genetics*, 57, 102661.

Fonnelop, A., Johannessenn, H., and Gill, P., 2015, Persistence and secondary transfer of DNA from previous users of equipment, *Forensic Science International: Genetics Supplement Series*, 5, e191–e192.

Fonnelop, A., Ramse, M., Egeland, T., et al., 2017, The implications of shedder status and background DNA on direct and secondary transfer in an attack scenario, *Forensic Science International: Genetics*, 29, 48–60.

Fitzgerald v The Queen [2014] High Court of Australia HCA 28.

Gans, J., 2014, *The DNA, the handshake and the didgeridoo: Fitzgerald v The Queen*, available at https://blogs.unimelb.edu.au/opinionsonhigh/2014/08/18

Gill, P., 2014, *Misleading DNA evidence*, Elsevier, New York.

Gill, P., 2019, DNA evidence and miscarriages of justice, *Forensic Science International*, 294, e1–e3.

Gill, P., Hicks, T., Butler, J., et al., 2020, DNA commission of the International society for forensic genetics: Assessing the value of forensic biological evidence – Guidelines highlighting the importance of propositions. Part II: Evaluation of biological traces considering activity level propositions, *Forensic Science International: Genetics*, 44, 1–13.

Goray, M., Fowler, S., Szkuta, B., et al., 2016a, Shedder status – An analysis of self and non-self DNA in multiple handprints deposited by the same individuals over time, *Forensic Science International: Genetics*, 23, 190–196.

Goray, M., Ballantyne, K., Szkuta, B., et al., 2016b, Commentary on Cale, C. et al., Could Secondary DNA transfer falsely place someone at the scene of a crime? *Journal of Forensic Sciences*, 61, 1396–1399.

Graham, E. and Rutty, G., 2008, Investigation into 'normal' background DNA on adult necks: Implications for DNA profiling of manual strangulation, *Journal of Forensic Sciences*, 53, 5, 1074–1082.

Helmus, J., Pfeifer, M., Feiner, L., et al., 2019, Unintentional effects of cleaning a crime scene—When the sponge becomes an accomplice in DNA transfer, *International Journal of Legal Medicine*, 133, 759–765.

Helmus, J., Poetsch, J., Pfeifer, M., et al., 2020, Cleaning a crime scene 2.0—What to do with the bloody knife after the crime? *International Journal of Legal Medicine*, 134, 171–175.

Jansson, L., Swensson, M., Gifvars, E., et al., 2022, Individual shedder status and the origin of touch DNA, *Forensic Science International: Genetics*, 56, 102626.

Johannessen, H., Gill, P., Roseth, A., et al., 2021, Determination of shedder status: A comparison of two methods involving cell counting in fingerprints and the DNA analysis of handheld tubes, *Forensic Science International: Genetics*, 53, 102541.

Kamphausen, T., Schadendorf, D., von Wurmb-Shwark, N., et al., 2012, Good shedder or bad shedder – The influence of skin diseases on forensic DNA analysis from epithelial abrasions, *International Journal of Legal Medicine*, 126, 1, 179–183.

Kamphausen, T., Fandel, S., Gutmann, T., et. al., 2015, Everything clean? Transfer of DNA traces between textiles in the washtub, *International Journal of Legal Medicine*, 129, 4, 708–714.

Kokshoorn, B., Aarts, B., Ansell, R., et al., 2016, Commentary on Cale C. et al., Could secondary DNA transfer falsely place someone at the scene of a crime? *Journal of Forensic Sciences*, 61, 5, 1401–1402.

Lee, L., Wong, H., Lee, J., et al., 2019, Persistence of DNA in the Singapore context, *International Journal of Legal Medicine*, 133, 1341–1349.

Lowe, A., Murray, C., Whitaker, J., et al., 2002, The propensity of individuals to deposit DNA and secondary transfer of low level DNA from individuals to inert surfaces. *Forensic Science International*, 129, 25–34.

Magee, A., Breathnach, M., Doak, S., et al., 2018, Wearer and non-wearer DNA on the collars and cuffs of upper garments of clothing, *Forensic Science International: Genetics*, 34, 152–161.

Manoli, P., Antonioni, A., Bashiardes, E., et al., 2016, Sex-specific age association with primary DNA transfer, *International Journal of Legal Medicine*, 130, 1, 103–112.

McKoll, D., Harvey, M., and van Oorschot, R., 2017, DNA transfer by different parts of a hand, *Forensic Science International: Genetics Supplement Series*, 6, e29–e31.

Meakin, G. and Jamieson, A., 2013, DNA transfer: Review and implications for casework, *Forensic Science International: Genetics*, 7, 434–443.

Meakin, G., Butcher, E., van Oorschot, R., et al., 2017, Trace DNA evidence dynamics: An investigation into the deposition and persistence of directly and indirectly transferred DNA on regularly used knives, *Forensic Science International: Genetics*, 29, 38–47.

Meakin, G., Kokshoorn, B., van Oorschot, R., et al., 2021, Evaluating forensic DNA evidence: Connecting the dots, *WIREs Forensic Science*, 3, 4, July/August, e1404.

Miller, M., Philpott, M., Olsen, A., et al., 2021, Technical note: Survey of extracellular and cell-pellet-associated DNA from 'touch' samples, *Forensic Science International*, 318, 110556–110557.

Nicas, M. and Best, D., 2008, A study quantifying the hand-to-face contact rate and its potential application to predicting respiratory tract infection, *Journal Occupational Environmental Hygiene*, 5, 347–352.

Noel, S., Lagace, K., Rogic, A., et al., 2016, DNA transfer during laundering may yield complete genetic profiles, *Forensic Science International: Genetics*, 23, 240–247.

Phipps, M. and Petricevic, S., 2007, The tendency of individuals to transfer DNA to handled items, *Forensic Science International*, 168, 162–168.

Poetsch, M., Markwerth, P., Konrad, H., et al., 2022, About the influence of environmental factors on the persistence of DNA – A long term study, *International Journal of Legal Medicine*, 136, 687–693.

Quinones, I. and Daniel, B., 2012, Cell free DNA as a component of forensic evidence recovered from touched surfaces, *Forensic Science International: Genetics*, 6, 1, 26–30.

R v Sumner, R v Fitzgerald [2013] SASCFC 82.

Raymond, J., van Oorschot, R., Walsh, P., et al., 2009, Trace evidence characteristics of DNA: A preliminary investigation of the persistence of DNA at crime scenes, *Forensic Science International: Genetics*, 4, 26–33.

Reither, J., Gray, E., Durdle, A., et al., 2021, Investigation into the prevalence of background DNA on flooring within houses and its transfer to a contacting surface, *Forensic Science International*, 318, 110563.

Samie, L., Hicks, T., Castella, V., et al., 2016, Stabbing simulations and DNA transfer, *Forensic Science International: Genetics*, 22, 73–80.

Samie, L., Taroni, F., and Champod, C., 2020, Estimating the quantity of transferred DNA in primary and secondary transfers, *Science and Justice*, 60, 2, 128–135.

Smith, P.A., 2016, When DNA implicates the innocent, *Scientific American*, 314, 6, 11–12.

Stella, C., Mitchell, J., and van Oorschot, R., 2017, Hand activities during robberies – Relevance to consideration of DNA transfer and detection, *Forensic Science International: Genetics Supplement Series*, 6, e3–e5.

Szkuta, B., Ballantyne, K., Koshoorn, B., 2018, Transfer and persistence of non-self DNA on hands over time: Using empirical data to evaluate DNA evidence given activity level propositions, *Forensic Science International: Genetics*, 33, 84–97.

Szkuta, B., Ansell, R., Boiso, L., et al., 2019, Assessment of the transfer, persistence, prevalence and recovery of DNA traces from clothing: An inter-laboratory study on worn upper garments, *Forensic Science International Genetics*, 42, 56–68.

Szkuta, B, Ansell, R., Boiso, L., et al., 2020, DNA transfer to worn upper garments during different activities and contacts: An inter-laboratory study, *Forensic Science International: Genetics*, 46, 102268.

Taupin, J. and Cwiklik, C., 2010, *Scientific protocols for the forensic examination of clothing*, CRC Press, Boca Raton, Florida.

van den Berge, M., Oxcanhan, G., Zijlstra, S., et al., 2016, Prevalence of human cell material: DNA and RNA profiling of public and private objects and after activity scenarios, *Forensic Science International: Genetics*, 21, 81–89.

van Oorschot, R. and Jones, M., 1997, DNA fingerprints from fingerprints, *Nature*, 387, 767.

van Oorschot, R., Glavich, G., and Mitchell, J., 2014, Prevalence of DNA deposited by the original user on objects after subsequent use by a second person, *Forensic Science International: Genetics*, 8, 1, 219–225.

van Oorschot, R., McColl, D., and Alderton, J., 2015, Activities between activities of focus – relevant when assessing DNA transfer probabilities, *Forensic Science International: Genetics Supplement Series*, 5, e75–e77.

van Oorshot, R., Szkuta, B., Meakin, G., et al., 2019, DNA transfer in forensic science: A review, *Forensic Science International: Genetics*, 28, 140–166.

Wickenheiser, R., 2002, Trace DNA: A review, discussion of theory and the application of the transfer of trace quantities of DNA through skin contact, *Journal of Forensic Sciences*, 47, 3, 442–450.

Worth, K., 2018, *Framed for his own murder*, The Marshall Project, April 19, available at http://www.themarshallproject.org

Zoppis, S., Muciaccia, B., D'Alessio, A., et al., 2014, DNA fingerprinting secondary transfer from different skin areas: Morphological and genetic studies, *Forensic Science International: Genetics*, 11, 137–143.

Chapter 5

Medical Exhibits

BOX 5

- Medical exhibit considerations.
- Sexual assault investigation kits – female genital and male genital swabs.
- Fingernail and skin swabs.
- 'Abandoned' material.

5.0 Introduction

Medical exhibits are collected from the body of a person to determine the presence of foreign DNA assumed to have been *directly* transferred from another person. If the DNA profiles obtained can be associated with sperm, semen or saliva then this has more meaning in sexual assault cases.

Close physical contact or the exchange of body fluids is presumed during sexual assaults. Traces to large amounts of DNA-containing material can remain on victims and perpetrators (Fonnelop et al., 2019).

DOI: 10.4324/9781003158844-5

The source of the body fluid and DNA are often in question if the DNA profile is purported to come from epithelial cells (skin or vaginal mucosa). This may be an issue where, for example, the DNA of a complainant is detected on a penile swab from the suspect. The prosecution may allege that a rape has occurred while the defence may claim that only social contact occurred.

The risk of indirect transfer has traditionally been considered as a result of flawed collection and analysis practices. Recently, indirect transfer has been considered in terms of non-intimate social contact, where the participants live in a domestic situation, may be known to each other or met at a social event. Bayesian analysis has been used to analyze the DNA results in terms of 'activities'. An assessment of the relation of the DNA obtained to a purported body fluid or cell is required.

Medical exhibits have different considerations regarding transfer, persistence and recovery compared to routine exhibits collected from crime scenes or clothing from a person. Due to the private nature of medical exhibits there are special requirements in collection, handling and storage. The persistence of DNA material is also very much dependent on the time elapsed since the alleged offence due to the natural movement of the body resulting in loss, bacterial degradation and drainage, and other effects such as washing.

Medical genital swabs from females are notable for their occurrence in miscarriages of justice and/or flawed initial investigations. This may be due to the fact that DNA from genital swabs purported to derive from sperm or semen has been wrongly associated to a person committing rape – see Farah Jama (Case 1.3), Kevin Brown (Case 2.5) and Adam Scott (Case 5.1, below). Flawed collection, handling and/or forensic analysis techniques contributed to a general misunderstanding and also a tendency for the DNA evidence to override any neutral or exculpatory evidence in these cases. This fallacy has been defined as the 'naïve investigator effect' by Gill (2014). The problem is that these are genital (intimate) swabs and thus there is a tendency to associate any DNA profile obtained to be derived from a criminal act or a 'sexual activity'.

The above cases also had 'sole-plank' DNA evidence or what is known as 'DNA-only' cases.

The 'compounded error effect' has also been described by Gill (2014) that can equally be ascribed to the above three cases. This effect combines the three main effects of the association fallacy (Chapter 2), the 'hidden perpetrator effect' and the 'naïve investigator effect'. Appendix 2 summarizes these fallacies and effects.

Strangely, DNA evidence from internal vaginal swabs pointed away from the accused in the Juan Rivera Case (Case 2.6), yet the investigators explained this away by bizarre theories. This may be a type of cognitive bias (or more specifically cognitive dissonance).

The evidence that is collected by medical personnel from the body of a person has particular requirements in collection and sampling as well as preserving. There are ethical considerations regarding consent for living individuals and the requirement for minimal discomfort and treating the person with respect and dignity. These extra considerations can be contrasted with collection of essentially abandoned material at a crime scene.

A complainant may have different requirements according to a jurisdiction regarding samples taken, in contrast to a suspect or an accused in a criminal case. A complainant may agree to certain samples or none at all; indeed they may not come forward with a complaint which has been viewed as an issue in ascertaining the prevalence of sexual assaults in a particular jurisdiction. Conversely, legislation may compel a suspect to provide medical samples under court order.

An autopsy of a deceased person has an implicit consideration of consent, as it is the aim of the jurisdiction to determine cause of death when a body is sent for autopsy. Evidence may be gathered from the body during the autopsy and sent for specialist examination.

5.1 Sexual Assault Investigation Kits

5.1.0 Introduction

Medical examination kits from living and deceased persons suspected of being the victim of a sexual assault generally comprise medical swabs taken from the internal orifices. Sometimes the external genital areas are also swabbed – such as the vulva or peri-anal regions of the body. Other skin regions may be swabbed, depending on the case information, such as the breast or chest area, or neck area.

Reference DNA samples are now usually taken from the mouth (buccal swabs) unless there is an allegation of oral rape whereby blood samples may be taken. Blood samples are also taken for reference DNA profiles at autopsy. 'Cold cases' may also have had blood samples taken as the reference, since previously blood samples were the routine when DNA analysis was not as sensitive as today.

A standard Sexual Assault Evidence Kit (SAEK) is used by the examining doctors and nurses to collect oral, rectal and vaginal swabs in Canada (Dziak et al., 2011). At the time of medical examination, microscope slide smears are prepared from each swab type. Following submission of the SAEK items to the forensic laboratory, microscopic examination of the smears for spermatozoa and chemical analysis for seminal acid phosphatase and prostatic specific antigen (p30) on the internal swabs is used to screen for semen. Depending on the results of the above tests, microscopic analysis may be conducted on slides made from internal swabs in the laboratory.

A medical officer may often smear the swab onto a microscope slide so that the slide can be examined for spermatozoa in the forensic laboratory, leaving the swab intact. This may improve efficiency for the forensic scientist but creates potential for error, loss and contamination because the evidence consists of two separate entities. Major English laboratories create their own microscope slides from swabs to ensure a direct connection between sperm detected and the swab and also retain a representative sample for microscopy (author information).

It is important to remember that the forensic biologist examines the swabs and/or slides taken from the person and thus there may be an assumption of a direct link between the sample on the swab and the body area from where it was collected. This assumption means there is no intervening material or objects that may introduce contamination, and/or no flawed collection practices, which may not necessarily be true. As the analysis of samples for DNA becomes ever more sensitive, collection, handling and forensic examination practices need to be ever more quality conscious.

5.1.1 Female Genital Medical Samples: Vaginal and Vulval Swabs

Traditionally the collection of medical exhibits in cases of alleged rape or sexual assault has focused on the collection of specimens from either the female or male complainant. Male DNA found in and on the vagina post-intercourse could be derived from spermatozoa within the semen ejaculate, male epithelial cells in the seminal fluid and/or biological material (e.g., skin-derived or saliva-derived DNA) on the penis, fingers, tongue or object that entered the vagina (van Oorschot et al., 2019).

Generally, seminal fluid survivability is less than that of sperm cells due to their protein nature (proteins will be broken down more quickly than sperm cells). It has also been reported that sperm cell survival time is longer in body orifices post-mortem. The environmental conditions of the body will have a major impact in determining survival times. Differences in sensitivity of semen and sperm cell detection assays will result in variations on reported survivability of the constituents of seminal fluid.

It is also known that seminal fluid drainage can occur, resulting in the transfer of semen from one location to another (e.g., drainage from the vaginal area to the rectal area or drainage from vagina to the underwear). This seminal fluid drainage can result in the loss of detection from its initial area.

5.1.2 Spermatozoa Presence and Persistence in the Vagina

The persistence of spermatozoa in the vaginal cavity in the forensic literature is often equated to its survival in the cavity. Much of the literature on survival times is decades old and performed on casework samples where the ground truth is unknown. This is especially an issue in cases of deceased persons.

Research on the persistence of foreign DNA in the internal and external vaginal areas of a female individual has until recently focused on the survival of spermatozoa (sperm). Studies have shown that time since intercourse has a direct effect on the ability to recover sperm from the female genital tract as sperm is destroyed relatively quickly in the hostile environment of the vagina. Sperm survival rates also decrease upon exposure to the environment. The ability to obtain a DNA profile from the semen donor in the vagina using ‘routine’ DNA profiling decreases rapidly after 24 to 36 hours post-coitus, but Y-STR profiling has increased the time interval. This correlates to studies on the persistence of spermatozoa in the cervix up to 7 to 10 days post-coitus (Mayntz-Press et al., 2008).

It has occasionally been found that spermatozoa survive for longer than one week in the deceased vaginal cavity.

Sperm loss after intercourse is due to vaginal drainage and the normal sperm degrading changes that occur over time, however studies have shown that phase of menstrual cycle is not a crucial factor. Loss can also occur during the multiple manipulations required of the differential separation used to separate the sperm from the non-sperm DNA fractions within the laboratory (van Oorschot et al., 2019).

A review from Canada (Dziak et al., 2011) examined time since intercourse, abbreviated as ‘TSI’, which refers to the post-coital interval between semen deposition and time of sample collection. Time since intercourse estimates based on the persistence of semen are limited by a number of factors, including variation in the amount of semen deposited, the sperm count and/or seminal acid phosphatase and P30 of the semen donor, activity of the complainant during the time period between intercourse and sampling and methods of sample collection, processing and storage. An Irish study noted that seminal acid phosphatase (AP) reaction times are unsafe and unreliable for the prediction of sperm on intimate swabs (Casey et al., 2017). Instead, and in this text author’s opinion, this test should be used on clothing and bedding for screening purposes to locate potential semen stains for further testing.

According to the Canadian study the manner in which vaginal swabs are processed removes the tails on intact spermatozoa (Dziak et al., 2011). As a result, when giving time since intercourse estimates, there is a distinction between vaginal smears and swabs. The maximum reported time frame for detecting *intact* spermatozoa in the vaginal cavity ranges from 18 to 72 hours.

A study from sexual assault cases was retrospectively performed to ascertain time since intercourse intervals at the

Forensic Science Ireland Laboratory for cases from over a period of 7 years (Casey et al., 2017). Note that these cases are not 'ground truth'. The particular laboratory prepares smears for microscopic examination from the swabs, instead of receiving smears from the medical examination. Internal vaginal swabs were found to have the longest time period since intercourse for survival of spermatozoa, typically up to 72 hours (ibid.). No sperm was recorded on any vaginal swab type beyond 96 hours.

There are insufficient studies in the literature regarding variables affecting removal of tails from spermatozoa due to forensic analysis, in this author's opinion. The notation of a single sperm with a tail – in addition to sperm heads – on a microscope slide in Case 1.3 (Farah Jama) was supposedly from a dried semen ejaculate in head hair from the first complainant, and thus not in liquid form when deposited. Further, it is poor logic to estimate time since deposition of sperm from the appearance of the sperm when there are multiple affecting variables.

The following case is a known miscarriage of justice that involved contamination in the laboratory, and an association error of the DNA profile statistic to a biological fluid (source) and then activity (sexual intercourse), that ultimately resulted in offence level (guilt or innocence). Direct transfer from the accused in the form of semen ejaculate was inferred by the laboratory to the vulva area of the complainant.

CASE 5.1 ADAM SCOTT (GILL, 2014; RENNISON, 2012)

A young man from the south of England was accused of raping a woman in Manchester in 2011, a city in the north to which he said he had never been. Adam Scott subsequently spent 5 months on remand in custody after it was allegedly found, through a database search, that his DNA profile matched that of DNA from semen found on a medical sample from the woman. The DNA was the sole evidence against the accused. He was released in March 2012 after being found 'the innocent victim of an avoidable contamination'.

Included in the exhibits of the case were six swabs from the medical examination of the complainant – two vulval, two low vaginal and two high vaginal swabs. Semen/sperm was detected on each of the swabs and separated from other cellular matter using differential extraction. Each fraction was analyzed for a DNA profile. The low, high vaginal and one vulval swab produced DNA profiles from the sperm fraction identified as those of the victim's boyfriend. The remaining vulval swab produced a mixed DNA profile from the sperm fraction containing the victim's boyfriend, and another unknown male with 17 out of 20 alleles present.

This unknown male DNA profile was uploaded onto the national DNA database. There was a partial DNA profile match of Adam Scott with a probability of 'one in one billion' chance of obtaining the DNA components in an unrelated person. The opinion of the scientist was that the DNA matching Scott most likely came from semen.

The scientific report stated:

Interpretation and conclusions: The DNA detected in the sample recovered from (victim's name) vulval swab can be accounted for by a mixture of DNA from (victim's boyfriend) and Adam Scott. In my opinion these findings are what I would expect if Adam Scott had some form of sexual activity with (victim's name).

In order to assess the overall findings in this case I have therefore considered the following propositions:

- *Adam Scott had vaginal intercourse with (victim's name)*
- *Adam Scott has never been to Manchester and does not know (victim's name)*

In my opinion, the scientific findings in relation to (victim's name) vulval swab provide strong scientific support for the view that Adam Scott had sexual intercourse with (victim's name) rather than he did not. However, given the position of the semen matching Adam Scott and an absence of semen on (victim's name) internal swabs, the findings do not specifically support vaginal penetration with ejaculation inside the vagina. They may also support vaginal–penile contact with external ejaculation or vaginal intercourse with no internal ejaculation. I have assessed the scientific findings based on the following scale of scientific support: no, weak, moderate, strong, very strong and extremely strong. Should any of this information change I may need to re-evaluate my findings. This is best done in advance of the trial.

Adam Scott was arrested on suspicion of rape and taken to Manchester where he was interviewed and denied being in Manchester at the time of the rape, or any involvement in the rape. The sole evidence against him was the partial DNA profile. It was established from cell site analysis of the telephone used by Adam Scott that the telephone was in Plymouth (the south of England) a few hours after the reported rape. The detective leading the investigation had concerns about the reliability of the DNA result and in the presence of contradictory, rather than supporting evidence, the forensic science laboratory was pressed and the case was reconsidered. Further profiling was done on the original DNA extract that contained the contaminated DNA and worked only to confirm the presence of Adam Scott's DNA. The reporting scientist sought advice from colleagues and the possibility of contamination was raised. This instigated a complete re-testing of the samples from the swabs that produced fresh results that showed Adam Scott's DNA was not present and an internal investigation commenced into the possibility of contamination.

The government inquiry found that a plastic tray sample holder was mistakenly re-used and loaded into equipment by a laboratory worker as part of the robotic DNA extraction process, instead of being disposed of into a rubbish bin. Saliva from Scott from an unconnected earlier 'spitting incident' was extracted in the same tray 'well', before DNA from the vulval swab from the queried case. Basic procedures for the disposal of plastic trays were not followed, records not maintained and nothing was done to mark trays once they had been used.

The Forensic Science Regulator stated it was unlikely that the case against Adam Scott would ever have proceeded to trial and in the absence of any further evidence the case would probably have been discontinued. However, this was of little comfort to Adam Scott who was remanded in custody for nearly 5 months.

The association error occurred when the scientist inferred that the DNA profile originated through sperm (source fallacy). Then again, when the DNA profile was used to infer that sexual intercourse took place, albeit the scientist stated the location (vulva) may support external ejaculation. The inference was that there was a direct transfer of ejaculate onto the vulva area of the body.

A transposition of the statistic from the DNA profile (sub-source) to semen (source) to sexual intercourse with ejaculation outside the vagina (activity) occurred in the above case. The movement through the hierarchy of propositions is fallacious reasoning. Further, the propositions by the scientist in their report addressed an alternative that was not within the scope of the biological evidence – 'Adam Scott has never been to Manchester'.

There are numerous problems when DNA evidence is used as the sole evidence in a criminal case, and there is also a considerable danger if the DNA evidence is inappropriately afforded greater weight than other evidence (Sense about Science, 2017). Note that the above case had only one swab out of six swabs with the DNA of interest, which should have brought additional attention (or red flags) considering the case in a holistic sense.

5.1.3 Post-mortem Medical Samples

Spermatozoa in internal medical samples from deceased persons can be detected (persist and be recovered) for significantly longer periods than for living individuals. Time frames have been reported up to 34 days (Dziak et al., 2011). A variety of factors may affect the persistence of spermatozoa including: seasonal temperatures, meteorological conditions, body positioning and shelter/storage conditions. Due to this variation, time since intercourse estimates for samples from living subjects are not applicable to samples taken post-mortem.

5.1.4 Y-STR Profiling

Autosomal or regular DNA profiling may not be undertaken if semen is not detected on intimate swab samples, due to the very low expectation of detecting male DNA suitable for meaningful comparison amid the high levels of background female DNA (Owers et al., 2018).

Y-STR typing may be used for the detection of male DNA in mixtures that contain an excess of female DNA (Gill et al., 2020). Y-STR profiling targets the Y chromosome only, and a female produces no results for this type of testing as they only have an X chromosome. One advantage of this testing is to isolate any male epithelial cells from the vast excess of female epithelial cells lining the vaginal tract. This has been relevant for analyses where a vaginal swab has seminal fluid detected but spermatozoa have not been identified.

Y-STR profiling analyses areas on the Y chromosome and these areas are physically linked and inherited as a unit called a 'haplotype'. The statistical interpretation of evidence from Y-STR profiling is challenging, and cannot be treated like autosomal (routine) DNA profiling where the areas analyzed (loci) are independent. It is important to consider whether the Y-STR profile has sufficient information to be used and whether it can be compared to a person of interest.

5.1.5 Persistence of Male DNA in the Vaginal Cavity

The persistence of male epithelial cells in the vaginal tract or in other female body orifices has not been considered until recently. There is an absence of literature research or empirical data regarding persistence of foreign DNA from epithelial cells after deposition inside the vaginal tract of a deceased person.

Recent literature discusses analysis of male DNA in cases where there has been suspected digital and/or penile penetration of the vaginal tract, and where no semen has been detected from vaginal swabs (Henry and Scandrett, 2019; Owers et al., 2018).

Guidelines in the United Kingdom for the post-incident time limit for recovering male DNA in vaginal swabs in digital penetration cases was increased from 12 hours to 48 hours in July 2014, influenced by increased sensitivity of techniques as supported by casework data (Owers et al., 2018). This time limit is less than that of spermatozoa (typically that of 3 days or more).

A study on the persistence of male DNA with no detected sperm after penile and/or digital penetration obtained Y-STR positive results up to 48 hours post-alleged offence (McDonald et al., 2015). However, as no samples were taken past 48 hours it was suggested that male DNA may be observed even later.

A South Australian study using casework samples where no semen was detected – in alleged cases where no semen was ejaculated – found full Y-STR profiles up to 44 hours after alleged deposition of male epithelial cells from penile and/or digital penetration (Henry and Scandrett, 2019). The study found the ability to generate an informative Y-filer Plus profile was not related to the time elapsed since the offence, the type of penetration or the anatomical origin of the sample (digital or penile). The study concluded that this indicates that the deposition and persistence of male epithelial cells does not necessarily follow a similar pattern to the deposition and persistence of semen and spermatozoa. The study also concluded that the reasons for this are unknown and warrant further study. The samples in the study were not taken under controlled conditions but casework samples dependent on casework information – which may not be verifiable.

It was stated in a paper by a United Kingdom laboratory that in situations where no semen is detected, it is often the case that any male DNA present will be at a low level, and so the potential for indirect transfer must be considered (Owers et al., 2018). Therefore, if a complainant and suspect are known to each other and they have had legitimate contact, this should be taken into account when assessing whether or not any male DNA matching the suspect could be the result of sexual activity rather than social contact and secondary transfer.

There is currently a lack of published literature in relation to the potential of external to internal transfer following contact with external vaginal surfaces, therefore this area would benefit from further research studies (Owers et al., 2018). This may be especially important in the medical examination and potential transfer of matter via examination tools and gloves, from material from the vulva area and other external surfaces.

A recent case from the files of this author illuminates potential issues in collection, handling and interpretation regarding Y-STR analysis of vaginal swabs from a child (author's case files).

CASE 5.2 CONTRADICTORY HIGH VAGINAL AND LOW VAGINAL DNA RESULTS

During the morning a 3-month-old baby girl was found deceased by her aunt. The mother had left her baby and her other child in the care of the accused and his partner at the home of her maternal grandmother. The accused and his partner were also caring for their own two children. The accused changed the nappy of the baby and generally tended to her. The accused was subsequently charged with murdering the baby as well as sexual penetration of the baby.

No spermatozoa (or semen) were detected from the vaginal, rectal or oral swabs from the baby. The vulva area was not examined. Y-STR profiling was performed (Y-filer plus) on the vaginal swabs.

The high vaginal swab had a partial, low-level Y-STR profile that indicated a mixture with only 8 sites (out of 27) with results reported. The report stated this was the same Y-STR profile as the accused and his son (*author note – a partial profile cannot be the same as a full reference profile*). The profile was expected to occur in 1 in 750 unrelated males in the general population.

The low vaginal swab had a Y-STR profile that was also partial and low level, and was stated to have the same Y-STR profile as the accused and his son except for a single discordance. The discordance was explained by a mutation for Y-STR profiling, or an artefact or the profile originating from someone other than the accused or his son. However, again, the Y-STR profile indicated a mixture.

Discordance at one site of a Y-STR profile with a reference sample cannot be explained by a mutation for Y-STR profiling, as these mutations on the Y chromosome occur during generational change. The different Y-STR profiles from the vagina (high and low) indicated problems in sampling and potential of contamination.

The potential for transfer from external areas to the internal cavities of the individual on examination is dependent on the method of swabbing and contamination mitigation practices in the examination rooms.

The sexual penetration charge was dropped and the accused pleaded guilty to manslaughter.

Increasing sensitivity of DNA analysis techniques demands an ever-increasing attention to collection and handling methods.

5.1.6 Background Levels of Male DNA in the Vagina

It is currently unknown whether male DNA may be present in samples collected from the vaginal cavity due to adventitious mechanisms, such as the shared use of bathroom facilities and washing machines (Albani et al., 2018). A study (ibid.) aimed to determine whether male DNA could be detected in samples collected from the vaginal cavity after everyday activities.

Samples from the vagina were subjected to DNA quantification and Y-STR amplification, with a finding of about 86% correspondences between the two (that is, not complete). In 29 samples (out of 300) there were no Y-STR profiles of three or more alleles despite detecting male DNA by quantification.

It was considered possible that trace amounts of foreign DNA may be collected from the labia, introitus or vaginal areas when an intimate swab is taken. Full Y-STR profiles were detected for up to 6 days post-intercourse (when this occurred); however time frames varied among individuals. It was thought that the use of hygiene products such as tampons may also remove some male DNA present in the vagina of menstruating women after intercourse (ibid.).

The high sensitivity of the Y-filer plus amplification system – meaning that full Y-STR profiles could be obtained from 112 pg of male DNA – emphasizes that extreme caution is paramount when interpreting low-level genetic data (ibid.). Fourteen per cent of the original samples displayed 1–2 unexpected allelic peaks (and two samples showed 3 allelic peaks) of which 71% did not correspond with the males in the study. There was a lack of reproducibility in low-level signals. The results at this low level indicate that it would not be prudent to form any kind of opinion about the likely origin of Y-STR results in such samples.

The relevance of one or two Y-STR alleles detected in a vaginal swab should therefore be evaluated with utmost caution in view of the framework of the case, following thorough internal validation and in addition to further testing on the sample (ibid.).

5.1.7 Vulval Swabs

The analysis of the vulva or areas external to the vagina is generally undertaken in a sexual assault medical examination – vulva swab areas are taken as part of the sexual assault kit. The analysis may be crucial if (a) external ejaculation is alleged and also no penetration and/or (b) there is an allegation of intimate touching without penetration.

5.1.7.1 Tests for saliva

Regarding alleged *oral sexual assault*, forensic biology tests for the presence of saliva in current practice such as the RSID saliva kits are non-specific when used on either vaginal/vulval swabs or the crotch from the underpants of the female complainant, due to cross-reactivity with vaginal secretions (Sari et al., 2019). Thus it is not currently possible using regular forensic techniques to ascertain saliva from a male (or indeed female) from vaginal areas/secretions. Cellular DNA from the male is the only indicator of deposition.

5.1.7.2 Saliva and buccal swabs

A corollary to the above is the analysis of exactly what is in samples purported to contain saliva, as compared to buccal swabs taken from scrapings inside the cheek – those that are often taken for DNA reference samples. It has been noted that buccal swabs and saliva swabs have been used interchangeably in the literature, despite evidence that both contain buccal cells and blood leukocytes (white blood cells) in different proportions (Theda et al., 2018). Buccal swabs contain a higher proportion of epithelial cells than saliva.

5.2 Rectal and Oral Cavities

The persistence of spermatozoa in the anal and oral cavities has been reported less frequently in the literature than for the vaginal cavity.

The rectal cavity may also be important in cases of suspected male/male anal rape.

The detection of semen in the anal cavity does not necessarily indicate that anal penetration has occurred, since semen may drain/transfer from the vaginal/ peri-anal area into the anal canal and possibly the rectum (Dziaz et al., 2011).

Given the rapid loss of semen from the oral and rectal cavities, the limited literature regarding intact spermatozoa persistence, and the lack of studies examining persistence of seminal acid phosphatase and the prostate specific antigen p30, such factors have not been used to further define the post-coital time interval. When semen is detected on oral and rectal samples, the maximum spermatozoa persistence times of 24 hours and 72 hours, respectively, should be utilized (ibid.).

The Irish study described previously (Casey et al., 2017) found it unlikely to detect sperm after 24 hours since intercourse from internal anal swabs, although one sample gave a result at 85 hours.

The persistence of sperm in the oral cavity decreases rapidly after 6 hours although it was reported the longest time recorded was 27 hours after intercourse (ibid.).

5.3 Male Intimate Swabs – Penile Swabs

5.3.1 Sampling Rationale

Internal and external genital swabs from in, and around, the orifices of male complainants may be examined in cases of male/male rape. This aspect is little researched in the forensic science arena especially regarding DNA transfer (this author opinion). Anal/rectal and oral swabs examined in female complainant studies may be used for information especially if there is DNA associated with spermatozoa that is foreign to the male complainant. Further research is recommended by this author.

The literature focuses on penile swabs taken in cases of alleged male/female rape. These cases also now include alleged penile penetration without ejaculation, as well as ejaculation, due to the increased sensitivity of DNA profiling techniques.

When no evidence is obtained from the body of the complainant in an alleged rape case the attention may turn to the body of the accused. Focus has gradually turned towards intimate body samples from the accused, particularly areas of the penis (shaft, coronal sulcus and glans). There is little published literature using controlled studies, postulated due to the difficulty of recruiting volunteer couples for such experiments (Bouzga et al., 2020).

Penile swabs from an alleged perpetrator were said to be processed as part of the routine procedure in Norway, but documentation on the expectation of finding female cells on post-coital penile swabs was scarce (Farmen et al., 2012). This raises the question of why such exhibits were taken. The Norwegian group aimed to investigate whether female cells can be found on penile swabs after vaginal intercourse. It was considered that, in principle, the perseverance of female epithelial cells on the exterior of the penis may provide useful information on the activity-level propositions when assessing a weight of evidence to scientific findings on penile swabs. The purpose of the study was to investigate whether female DNA can be found on penile swabs, sampled at time intervals from 0 to 24 hours after vaginal penetration (ibid.). The study found sufficient female DNA for analysis on all penile swabs collected between 0 and 24 hours after consenting vaginal–penile sexual intercourse.

5.3.2 Activity-level Propositions

A body fluid forum group in the United Kingdom and Ireland performed research to assist in cases where semen is absent on vaginal swabs, but female DNA is present on penile swabs or male underwear, and the issue to be addressed is whether or not sexual intercourse occurred (Jones et al., 2016). This may be especially relevant when the complainant and accused are known to each other and have been in social contact prior to the alleged event. This study aimed to investigate the frequency and amount of female DNA transferred to the penis and underwear of males following staged non-intimate social contact with females and simulated urination.

The initial trial swabbed the shaft of the penis, and DNA corresponding to the DNA profile of the female participant was detected on 4 of the 30 penile shaft samples at 0 hours (ibid.). During a 6-hour time delay trial in the above study the shaft, coronal sulcus and glans of each male volunteer's penis was swabbed. No matching female DNA was detected on any of the penile shaft, coronal sulcus or glans samples collected after the staged non-intimate social contact events. Note that less-sensitive DNA analysis techniques were used than today (28 cycles of amplification and a 3100 sequencer).

A Norwegian/Swiss study examined different situations regarding vaginal intercourse or secondary transfer of epithelial cells by skin contact and the deposition of DNA on the penis (Bouzga et al., 2020). They noted that none of today's routine methods or presumptive tests can decipher if the female DNA profile originates from skin or mucous epithelial cells from the vagina. The study hypothesized that vaginal mucosa epithelial cells will be transferred to a greater extent than skin epithelial cells in a secondary transfer scenario, as skin is considered to be a poor source of DNA compared to the vaginal mucous. The results showed that penile swabs following intercourse produce significantly higher DNA concentrations than after secondary transfer by skin contact. The study also concluded that it is possible that secondary transfer from a good shedder could be mistaken as direct transfer of mucous cells.

The study, however, examined a scenario where penile swabs were taken within 1 hour of the event and under hygiene conditions not concordant with crime situations (persistence and recovery for the vaginal fluid was also maximized). Further, 2 couples out of the 11 couple volunteers produced higher quantities of female DNA from the skin experiments than from the vaginal intercourse experiments, which was postulated as due to a high shedder status of the female (not tested).

Unfortunately the above study had in the title a reference to prediction of results, rather than an analysis of the results themselves, and has been critiqued (Biedermann, 2020). The prediction of whether female DNA is more likely to result from secondary transfer of skin cells or from vaginal penetration is commenting on the propositions or the scenarios, which is not the remit of the forensic scientist. The scientist can only evaluate results in the light of competing propositions, and not opine as to causes such as alleged modes of transfer. The Bouzga et al. paper stated in their conclusion that 'it is therefore possible to predict the scenario that lead to deposition of different cell types' but the main point here is that we do not predict the past (Biedermann, 2020). Statistical methods were also used that suggested the widespread but false idea that the answer to the question 'What is the cause of this observation?' could somehow be cranked out from scientific data alone (ibid.).

This discourse highlights problems in understanding the role of the forensic biologist and communicating to the forensic science community and ultimately the court (see Chapters 1 and 2 of this book). The term 'prediction' has been particularly criticized, and this is relevant for trace DNA amounts where the body source cannot be associated through serological tests. It has been said that the forensic science literature is replete with the use of the term, most notably in forensic genetics (Biedermann, 2020).

A method to detect epithelial cells derived from vaginal mucosa (as compared to cells from the skin) may assist in sexual assault cases, especially DNA obtained from penile swabs. Messenger RNA (mRNA) markers have been developed for a wide range of body fluids such as saliva, skin and vaginal mucosa, and more laboratories are implementing these markers (Johannessen et al., 2022). However, these markers may not be specific to the particular body origin area (i.e., vaginal mucosa may not be able to be differentiated from mucous from other body origins such as the nose).

Addressing activity-level propositions, such as intimate activity versus social contact, requires an assessment of probabilities of transfer, persistence, prevalence and recovery of DNA. A Bayesian analysis was performed for such a situation in relation to DNA and mRNA from penile swabs, fingernail swabs and boxer short clothing in the above paper (ibid.). The propositions were:

- Hp: the suspect had vaginal intercourse with the victim.
- Hd: the suspect and the victim only had social contact.

The DNA of the person of interest (POI) is present, with or without an unknown contributor. These propositions were designed to fit a general scenario often encountered in case work.

Nodes in the Bayesian network for Hp calculations related to the probability of direct transfer, persistence and recovery of the DNA of the POI. For Hd calculations the probability of the indirect transfer, persistence and recovery of the DNA of the POI were used. A positive finding of very strong supporting evidence corresponded to DNA results that support the proposition that the POI is a donor with a sub-source likelihood ratio of > 10,000. Positive results were detected in generally over 90% of the intimate samples depending on time since deposition and 33% of the non-intimate samples (ibid.). Indirect transfer was explained as the mechanism for the non-intimate samples. Full profiles of the POI (i.e., there were 46 alleles) were detected in some of the non-intimate contact penile samples. This highlights that the evaluation of DNA evidence must be carried out with care when family members or co-inhabitants are involved, as there are likely to be several sources of their DNA in the household that can be innocently transferred.

It was shown that the value of the evidence is mainly affected by the high DNA quantity that is recovered from the POI. The detection of vaginal mucosa had a low impact upon the resultant likelihood ratio. Time since intercourse and the amount of DNA are explanatory variables that may impact probability of detection of vaginal mucosa as the dependant variable. An important finding from the above study is that an understanding of transfer, persistence, prevalence and recovery of DNA and mRNA body fluid markers is necessary to decide an examination strategy based upon expectations of findings, given the alternative propositions.

5.4 Fingernails

5.4.1 Collection, Transfer and Persistence

Fingernail swabs and cuttings have routinely been taken for decades during the autopsy of a deceased. They are used as evidence in 'cold cases' from victims (not just deceased) because they may be one of the few surviving samples from the original investigation. However, such evidence can be stored in containers that may not be self-evidently important. They may be in jars not stored in freezers or in paper envelopes in case files.

A study using less-sensitive techniques than today (Dowlman et al., 2010) showed that good-quality DNA profiles from fingernails were associated with recent intimate contact with one or more others, such as partners, flat mates and/or children. Low-level DNA profiles were associated with all levels of contact.

Another study found that variable amounts of foreign DNA can be found in fingernail samples from co-habiting couples without any sexual contact having occurred (Malsom et al., 2009). The conclusion was that care must be taken when interpreting DNA profiles from individuals, as two of the high-level profiles in the study were from individuals sharing accommodation, but who had not had intimate contact 48 hours prior to sampling.

Samples are not usually taken from a suspect unless there is a relatively short time frame since the event. This interval appears to be according to the jurisdiction.

Fingernail scrapings and clippings may be routinely examined for the presence of foreign DNA profiles in forensic casework where the case history suggests their evidentiary relevance (Matte et al., 2012). DNA profiles from the donor, a foreign source or both may be obtained from debris collected beneath the fingernails of victims of violent crimes, possibly due to the transfer of skin and/or other body fluids from intimate physical contact. This study (ibid.) combined an analysis of 265 cases with additional experiments involving simulated scratching. They found foreign DNA 33% of the time, versus having found it 19% of the time in the general population. The scratching experiments produced 37% foreign DNA with 17% that was reportable. More vigorous scratching produced 30% reportable profiles. Normal activity over a 6-hour period reduced the percentage of foreign DNA considerably, from 33% to 7%.

Controlled experiments have showed that there is a very high probability of vaginal cells being transferred to fingernails after digital penetration (Flanagan and McAlister, 2011), but factors such as persistence and recovery affect the detection – such as time interval to collection and sampling techniques and analysis. A persistence study showed at 6 hours after digital penetration, full female profiles were still detected from all the samples, but some male DNA was also observed. As time increased since the penetration, the detection of female DNA also reduced. At 12 hours post-penetration, dish washing and hand washing were found to significantly reduce the detection of female DNA, and at 18 hours post-penetration, these factors, together with time since penetration were significant (ibid.).

If the complainant and the suspect have had undisputed recent social contact prior to the alleged offence, fingernail scrapings are not examined in Norway (Fonnelop et al., 2019).

Loss of DNA is promoted during daily activities (e.g., hand washing). If the victim is deceased, then the foreign DNA profile would not decay unless exposed to environmental and bacterial attack. An interesting New Zealand study from 2003 concerned the submersion of bodies in water and subsequent DNA analysis of fingernail samples (Harbison et al., 2003).

CASES 5.3 AND 5.4

Material was collected from under the fingernails of two women in two separate homicide cases. The first case had fingernail clippings from a 33-year-old female submerged in bath water fully clothed, and who had died from multiple blows to the head within the previous 2 hours of being found. Reference blood samples from the deceased and her family members were provided for comparison. Fingernail clippings from the left hand had a DNA profile with the female deceased as the minor contributor and the other male component was uploaded to the national DNA database. A DNA hit occurred to a male who was already a suspect, and this male was subsequently tried and convicted for homicide.

The second case of the above study (ibid.) concerned a 29-year-old female suicide victim and homicide suspect found floating in Auckland harbour. Approximately 3 hours earlier, she had been seen falling into the water from a motorway over a bridge whilst holding her 11-month-old baby. The deceased had been observed in an altercation with a male colleague about 1 hour prior to falling from the bridge, and she allegedly scratched the male's face. The fingernails of the deceased were clipped at the autopsy and supplied as left- and right-hand samples. Reference blood samples from the deceased and the male colleague were also supplied for comparison. The female victim was the minor component and the other component corresponded to the male colleague. No charges were laid.

The cases above used less-sensitive DNA profiling techniques than today. The paper (ibid.) noted that in cases of homicide the actions of the deceased during the incident are usually not fully known, so the mandatory collection of fingernails from all deceased regardless of the scenes that they are found in, including submersion in water, was recommended.

The study using routine DNA techniques and mRNA markers described above for penile swabs (Johannessen et al., 2022) noted positive findings of sub-source likelihood ratio greater than or equal to 10,000 for 50% of non-intimate social interactions for fingernail samples.

5.4.2 Association Fallacies

A case that is notable in the forensic literature due to controversy over expert opinions on DNA obtained from fingernails is the *R v. Weller* case from the United Kingdom. This case has also been used to demonstrate the association fallacy and also the nature of expert opinion (Gill, 2014). Contact between the victim and the defendant was not denied and there was some prior expectation that DNA transfer would occur between the two (ibid.). Digital penetration was alleged and the inference was that vaginal cells were present underneath the fingernails of the accused due to the relatively high quantity of DNA attributed to the complainant. The source of the DNA was inferred solely by its presence and circumstances of the case and no attempt was made to evaluate if vaginal cells were present – although this is still a difficult endeavour but the limitations should have been stated. Consequently this is a source association fallacy (also see Appendix 2).

CASE 5.5 *R V. WELLER* [2010] APPEAL

The appellant was convicted of sexual penetration of a 16-year-old girl in 2006. The complainant's evidence was that she went to a party where the appellant was present and she drank steadily until she vomited and was dizzy. She was helped to a bedroom by the appellant and got into bed as she was unwell. She stated that the appellant inserted his fingers into her vagina. Eventually she left the premises and met a friend, told him what had happened and he took her to the police station where she was seen at about 4 o'clock in the morning. She told the police that she had been indecently assaulted. A medical examination followed which confirmed injuries were consistent with the allegation.

The appellant's account of events was that he confirmed what she had said about him looking after her when she became sick. He said he had to pull the hair out of her eyes to stop her vomiting on it. He had helped her into bed. When in bed he had checked her several times and on one occasion had to put her into the recovery position. He had to pick up her clothes, including her underwear which she had left on the floor. He strongly denied the account of the sexual assault that she had given.

The fingernails of the appellant were swabbed – on the right hand there was only found a DNA profile for the appellant. On the left hand a full mixed profile was obtained. The major profile was that of the appellant but there was a minor full profile that matched the complainant; this was agreed by all parties. At the trial the sole issue was the strength of the evaluation. On the appeal the issue was whether there was sufficient scientific certainty and sufficient scientific reliability for an expert to be able to express an opinion on the evaluation of the possibilities.

The prosecution expert said in evaluating the possibilities that she believed that the most likely source of the DNA was the vagina. She went on to say that she considered that it provided strong scientific support for the view that the source of the DNA was contact with the vagina. The defence expert said that insertion into the vagina was a ready explanation for the DNA but the term 'strong' was setting it too high.

An expert at the appeal stated that if a proper review had been undertaken by either of the experts of the available scholarship and available papers and a proper analysis of the scientific evidence had been made, it would have shown that the state of scientific knowledge was such that on analysis on the facts of the case neither of those two experts could properly have given the evaluative evidence that they gave.

The appeal court dismissed the application. The DNA evidence was described as a 'makeweight' which proved conclusively that the jury was right in the verdict that they reached.

The court stated that experience was a sufficiently reliable scientific basis for a forensic science officer to give evidence of the evaluation of the possibilities of transfer in the circumstances of this case. Their long experience of dealing with expert witnesses in different fields is that experts often rely of necessity on unpublished papers and on their own experience and experiments. The court was entirely satisfied that there was no fresh evidence.

The finding regarding experience-only-based opinion has been criticized in the literature (for example Champod, 2013). Experience-only-based opinions cannot be challenged beyond the sterile opposition between mere opinions. Requiring the disclosure of structured data opens the route to a new type of debate regarding the relative merits of the assessments provided (ibid.).

A probabilistic assessment of the case circumstance was performed by Gill (2014). Using data published to that date, and in the absence of a test for vaginal cells, a likelihood ratio of 16 was obtained for the DNA evidence: the evidence is 16 times more likely if it originated from digital penetration than if it originated via some innocent source. That is, the strength of the evidence was weak and was many orders of magnitude lower than that required for a DNA profile to be considered probative (ibid.).

5.5 Areas of Skin

Forensic medical staff may take swabs (for the purposes of DNA analyses) from an area of skin other than hands or fingernails where physical contact with the offender is believed to have taken place, to assist in the potential identification of the offender. Current literature however is lacking in the value of such sample collection (Bowman et al., 2018).

The ability to successfully obtain an 'offender's profile' from the skin of the victim is complicated by normal non-self (or background) DNA.

One study (Fonnelop et al., 2019) retrospectively examining cases described a possibility of detecting male DNA from epithelial cells shed from the penis, hands or from saliva, and deposited during contact with other parts of the body, e.g., grabbing of the neck, breasts and kissing or biting the victim's skin. These areas can be sampled to look for the perpetrator's DNA. The persistence and detection of epithelial cells deposited on the body will be affected by factors such as new contacts, activity and personal hygiene, e.g., bathroom visits or showering (ibid.).

5.6 Abandoned Material

Exhibits collected from crime scenes or associated with them, and not obtained from the body, are considered abandoned. Due to the deposition of biological matter such as semen on various surfaces including clothing fabric, there is a requirement for an understanding of any additional substances that may produce false positives in testing for biological fluid source.

A bizarre case from the files of the author illustrates the many issues involved in such cases (not considering continuity) including transfer and multiple fallacies. There were similarities to the case of *R v. Weller* (Case 5.5) although the below matter was 5 years after that appeal and there was no material collected from the accused (apart from his reference sample).

CASE 5.6 SEMEN/SALIVA IN A RUBBISH BIN

A man was charged with oral sexual assault of a work colleague in a motel room. A group of work colleagues spent the night in a couple of motel rooms during an overnight working sales trip, and had a drinking session in the room shared by the complainant and another female work colleague. The complainant became drunk and was taken to bed by the accused as she was in an apparent state of sleep. She alleged that later the accused went in and out of her room and touched her inappropriately whilst the others were on the balcony of the room. She said he ejaculated inside her mouth while she was lying in bed. She spat it out in a rubbish bin by her bed. She collected the plastic bag lining the bin and containing the contents, took it home and submitted it to the police as evidence.

The bin had previously been in a toilet, and inside it there were beer bottles, tissues and napkins that had been used to wipe down surfaces in the room. The plastic bag was submitted to the forensic laboratory and contained apparent hairs (later noted by the court to be from a wig) and tissues (no bottles). The plastic bag was screened with an alternative light source and certain areas fluoresced. Biological material from all three samples was contended by the prosecution to likely contain saliva from the complainant and semen from the accused.

Samples from three of these areas, all from the inner surface of the plastic bag, tested positive for the seminal acid phosphatase (AP) screening test which is non-specific for semen. They also tested positive for a test for saliva (RSID saliva) which has false positives.

Spermatozoa were not detected on the three samples from the plastic bag. However, this did not deter the prosecution as they stated the accused had a vasectomy and thus did not produce sperm (information from his doctor). Y-STR technology was not used on the samples from the bag. Saliva was purported from the samples from the plastic bag using the RSID saliva test (which also has false positives).

A 'P30' test was performed on three samples from the plastic bag. This is a test for 'prostate specific antigen' which is found in semen. This test is known not to be specific to the prostate only and perhaps should be renamed 'prostate secreted antigen'. Other human secretions may react positively to this test as well as biological and chemical substances. During the trial the laboratory scientist stated 'in my opinion the P30 test in combination with the AP tests and the DNA profiling results confirms the presence of semen'.

The scientist later stated in testimony:

> 'In my opinion, the results of the biological testing are still more readily explained by the presence of semen, and also the DNA profile results are more readily explained by the presence of semen from a vasectomized male. But I don't discount the fact that there could be other substances there that have individually provided false positives to the biological tests'.

Later it was clarified: 'but not for both tests, AP and P30'.

Although no sperm was detected a differential extraction was performed on samples from the bag – 'sperm fraction' and 'non-sperm' fractions were DNA analyzed (note this is inappropriate terminology when no sperm has been detected). Further, these fractions were also called 'male fraction' and 'female fraction', respectively. The major DNA profiles from each of the 'non-sperm' fractions had a likelihood ratio of 100 billion compared to the complainant's reference DNA profile. In fact, there was a major female component to all three of the DNA profiles obtained with a trace male component (very much lower Y allele) and only a few additional alleles. Note further confusion in that it was postulated the presence of male cells in a 'female fraction', and it is not possible to separate male and female epithelial cells.

The values of the likelihood ratio for consideration of DNA of the accused contribution to the mixed DNA profiles from the 'non-sperm' fractions obtained varied from 140 (strong support) to 64 (moderate support) to 12,000 (very strong support). It is noted that 100 billion is considered as 'extremely strong support'.

The laboratory staff DNA likelihood ratio cut-off threshold was 10,000; that is, any likelihood ratios for staff DNA profiles less than 10,000 compared to evidentiary samples were considered to 'match by chance' – an adventitious match. This was put to the laboratory scientist who replied that calculations were performed for staff comprising 400 members (after receipt of the defence expert report) on the 'non-sperm' fractions. One staff member produced a likelihood ratio of 22 for sample 1, one produced a likelihood ratio of 532 for sample 2 and another produced a 7 for sample 3. The scientist explained that these likelihood ratios did not support the contribution of a staff member to the DNA profiles. Apart from confusion, this aspect serves to illustrate the non-specificity of the trace components of the three DNA profiles.

The following exchange occurred towards the end of cross-examination in the trial between the defence barrister and the laboratory scientist:

'There is nothing to say that this DNA could not have come from things like skin cells, saliva or spittle; is there? – So you want me to consider the DNA profiles –

By itself? – in the absence of the biological testing? I'm not sure I can do that, because I know the results of the biological testing.

I've nothing further, Your Honor'.

The accused was convicted at trial of one count of indecent assault and three counts of rape and sentenced with a 7-year prison term. There was an appeal on other matters but this was dismissed.

There was no testing on areas that did not 'fluoresce' so it was unknown whether there were substances that produced reaction to the tests for semen. The association of a minor trace DNA component of a DNA profile to semen was an association error although couched in terms of 'my opinion'. Presumably the minor male DNA component was purported to come from cellular material in semen (although never stated) and not spermatozoa because of the vasectomy. The trace amounts of the DNA associated with a male could not be associated with cells from semen as opposed to any other source. Further fallacies regarding bias might occur to any reader of this case.

The above trial occurred about 7 years ago and it is hoped that such DNA pronouncements would not occur today, although this is not assured.

REFERENCES

Albani, P., Patel, J., and Fleming, R., 2018, Background levels of male DNA in the vaginal cavity, *Forensic Science International: Genetics*, 33, 110–116.

Biedermann, A., 2020, Letter to the editor: Commentary on 'Is it possible to predict the origin of epithelial cells? – A comparison of secondary transfer of skin epithelial cells versus vaginal mucous membrane cells by direct contact', *Science and Justice*, 60, 201–203.

Bowman, Z., Mosse, K., Sungalla, A., et al., 2018, Detection of offender DNA following skin-to-skin contact with a victim, *Forensic Science International: Genetics*, 37, 252–259.

Bouzga, M., Dorum, G., Gunderson, K., et al., 2020, Is it possible to predict the origin of epithelial cells?- A comparison of secondary transfer of skin epithelial cells versus vaginal mucous membrane cells by direct contact, *Science and Justice*, 60, 3, 234–242.

Casey, D., Domijan, K., MacNeill, S., et al., 2017, The persistence of sperm and the development of time since intercourse (TSI) guidelines in sexual assault cases at forensic science Ireland, Dublin, Ireland, *Journal of Forensic Sciences*, 62, 3, 585–592.

Champod, C., 2013, DNA transfer: Informed judgement or mere guesswork? *Frontiers in Genetics*, 4, 3000.

Dowlman, E.A., Martin, N.C., Foy, M.J., et al., 2010. The prevalence of mixed DNA profiles in fingernail swabs, *Science and Justice*, 50, 2, 64–71.

Dziak, B., Parker, L., Collins, V., et al., 2011, Providing evidence based opinions on time since intercourse (TSI) based on body fluid testing results of internal samples, *Canadian Society of Forensic Science Journal*, 44, 2, 59–64.

Farmen, R., Haukeli, I., Ruoff, P., et al., 2012, Assessing the presence of female DNA on post-coital penile swabs: Relevance to the investigation of sexual assault, *Journal of Forensic and Legal Medicine*, 19, 7, 386–389.

Flanagan, N. and McAlister, C., 2011, The transfer and persistence of DNA under the fingernails following digital penetration of the vagina, *Forensic Science International: Genetics*, 5. 479–483.

Fonnelop, A., Johannessen, H., Heen, G., et al., 2019, A retrospective study on the transfer, persistence and recovery of sperm and epithelial cells in samples collected in sexual assault casework, *Forensic Science International: Genetics*, 43, 102153.

Gill, P., 2014, *Misleading DNA evidence: Reasons for Miscarriage of Justice*, Academic Press, Elsevier, London and New York.

Gill, P., Brenner, C., Brinkmann, B., et al., 2020, DNA Commission of the International Society of Forensic Genetics (ISFG): Recommendations on the interpretation of Y-STR results in forensic analysis, 2020, *Forensic Science International: Genetics*, 48, 102308.

Harbison, S., Petricevic, S., and Vintiner, S., 2003, The persistence of DNA under fingernails following submersion in water, *International Congress Series*, 1239, 809–813.

Henry, J. and Scandrett, L., 2019, Assessment of the Yfiler® Plus PCR amplification kit for the detection of male DNA in semen-negative sexual assault cases, *Science and Justice*, 59, 480–485.

Johannessen, H., Gill, P., Shanthan, G, et al., 2022, Transfer, persistence and recovery of DNA and mRNA vaginal mucosa markers after intimate and social contact with Bayesian network analysis for activity level reporting, *Forensic Science International: Genetics*, 60, 102750.

Jones, S., Scott, K., Lewis, J., et al., 2016, DNA transfer through non-intimate social contact, *Science and Justice*, 56, 90–95.

Malsom, S., Flanagan, N., McAlister, C., et al., 2009, The prevalence of mixed DNA profiles in fingernail samples taken from couples who co-habit using autosomal and Y-STRs, *Forensic Science International: Genetics*, 3, 57–62.

Matte, M.L., Williams, L.R., Frappier, R., et. al., 2012, Prevalence and persistence of foreign DNA beneath fingernails, *Forensic Science International Genetics*, 6, 2, 236neti.

Mayntz-Press, K., Sims, L., Hall, A., et al., 2008, Y-STR profiling in extended interval (≥ 3 days) Postcoital cervicovaginal samples, *Journal Forensic Sciences*, 53,2, 342–348,

McDonald, A., Jones, E., Lewis, J., et al., 2015, Y-STR analysis of digital and/or penile penetration cases with no detected spermatozoa, *Forensic Science International: Genetics*, 15, 84–89.

Owers, R., McDonald, A., Montgomerie, H., et al., 2018, A casework study comparing success rates and expectations of detecting male DNA using two different Y-STR multiplexes on vaginal swabs in sexual assault investigations where no semen has been detected, *Forensic Science International: Genetics*, 37, 1–5.

R v Weller Appeal [2010] EWCA Crim 1085 Royal Courts of Justice, London.

Rennison, A., 2012, Report into the circumstances of a complaint received from the greater Manchester police, FSR-R-618, *Forensic Science Regulator*, September 17.

Sari, D., Hitchcock, C., Collins, S., et al., 2019, Amylase testing on intimate samples from pre-pubescent, post-pubescent and post-menopausal females: Implications for forensic casework in sexual assault allegations, *Australian Journal of Forensic Sciences*, 52, 6, 618–625.–

Sense About Science, 2017, *Making sense about forensic genetics*, available at http://www.senseaboutscience.org

Theda, C., Hwang, S., Czaigo, A., et al., 2018, Quantitation of the cellular content of saliva and buccal swab samples, *Scientific Reports Nature*, 8, 6944.

van Oorschot, R., Szkuta, B., Meakin, G., et al., 2019, DNA transfer in forensic science: A review, *Forensic Science International: Genetics*, 38, 140–166.

Chapter 6
Clothing and Implements

BOX 6

- Clothing.
- Knives.
- Firearms.
- Equipment and tools.
- Communal items and spaces.

6.0 Introduction

This chapter discusses 'personal' exhibits that are worn, used as weapons or tools or other personal items that may occur as exhibits in criminal cases, and also their environs such as rooms in a house. Participants in crime events are usually clothed (although not always, notably sexual offences) and clothing directly involved. Instruments or tools may be used as weapons and implements, knives and firearms especially so. The items discussed in this chapter are portable and may move from place to scene/s, gathering and removing DNA along the way. Non-self DNA on these

DOI: 10.4324/9781003158844-6

items is of interest, as well as the level of background DNA and owner or user DNA on the exhibit.

As described earlier in this book, DNA that can be readily associated to a specific biological fluid or matter such as semen or blood with relatively large quantities is a simpler evaluation in the meaning of the case. Semen and blood are associated with sexual activity or violence. Just 'DNA' may be associated with daily living. The items discussed in this chapter have predominantly 'trace DNA' as the consideration, where a person may be proposed to 'hold' a weapon, or 'grab' a pair of underpants to remove them as alleged in the case.

The evaluation of trace DNA in these situations is recommended by many in the literature and forensic organizations to be performed at 'activity' level, and not just 'sub-source' level (not only the DNA profile). This may address the question of 'How did the DNA get there?' The deposit on a handle of a knife may have a mixed DNA profile of three people, but that does not imply that any of them (and certainly not all at once) stabbed someone, held up a bank or store or performed some other criminal activity with the knife. Evaluations using 'activity level' propositions are recommended in these situations (e.g., Gill et al., 2020; Willis et al., 2015). Even so, more empirical data and more robust experiments than so far described in the literature are advocated to assist in any evaluation.

An important issue is the level of 'background DNA' or 'pre-existing' DNA present on a garment and personal items due to everyday interactions, when a garment or item is worn or item is used and contacts surfaces and/or people – including non-self DNA. Self and non-self DNA may be readily available for transfer and deposit on other surfaces, including clothing and personal items of other persons.

The studies described in this book show that the DNA of family members and co-habiting partners, co-residents, work colleagues and acquaintances and 'unknowns' may be detected on personal items such as clothing, equipment and their environs of a person of interest. Thus during forensic examinations reference samples of 'known persons' may be required to assist interpretation of not only mixture DNA profiles but also single-source and partial DNA profiles obtained from exhibits.

Many studies describe that it is possible for the DNA of an original handler/wearer of an item to persist for lengthy durations after being used by a second person. However, there are differences in outcomes depending on the material and composition of the item, various activities involved and the duration of those activities as well as the examination and interpretation methods used. This is the issue pointed out in Chapters 1 and 3 of this book (e.g., Gosch and Courts, 2019; van Oorschot et al., 2019).

Examination techniques are not only confined to the forensic DNA laboratory but also to any personnel examining the exhibit for the first time. These personnel may not necessarily have considerations in mind desired by DNA examiners – who may subsequently want to use 'activity level' assessments in their interpretations. Exhibits may be screened by police personnel and sampled for visible deposits, perhaps using some of the valuable samples in screening tests. The DNA examiner may not be aware of the original dimension of any deposit, the screening tests used or even the type of sampling method employed. A Norwegian study on the transfer of DNA during examination described steps to reduce the risk of contamination including 'evaluate the necessity of police pre-examining an exhibit prior to DNA analysis' (Fonnelop et al., 2016).

When interpreting the case relating to the activities involved there is a need to apply data from scenarios most closely aligned with the conditions of the case (van Oorschot et al., 2019). Further, as the origin of the DNA from hands, for example, is often unknown there may be conflicting data (Burrill et al., 2018). Even our current understanding of the content of the so-called 'touch' DNA deposits and the origins of potential DNA is limited, as it is unclear precisely what is being 'shed' (see Chapter 4).

Indeed, it may be that the examiner cannot report any finding in relation to an 'activity' level, or source level, and report only on sub-source (the DNA profile) with the appropriate limitations and caveats about what this means – and does not mean.

6.1 Clothing

6.1.1 'Wearer' and 'Toucher' DNA

The finding of DNA other than that of the purported wearer of a garment has resulted in criminal cases proceeding on this basis. The premise is that there is 'non-self' DNA on the garment that has originated during the criminal activity, which may not necessarily be the case.

A study in 2016 noted that there was insufficient literature in relation to DNA interpretation issues when investigating who wore rather than touched a garment (Breathnach et al., 2016). Male underpants worn for more than 12 hours were grabbed on the waistband by female volunteers. The study showed that the 'wearer' could be detected in just over half of samples and the 'toucher' was detected in just over 11% of samples. Reportable background DNA (non-wearer and non-toucher) occurred in 14% of samples and may affect the assessment of who wore the garment.

A Norwegian paper examined an 'attack' scenario in which an 'attacker' grabbed the shoulder area of a T-shirt in a controlled office environment (Fonnelop et al., 2017). Background DNA found on samples from the shirts comprised multiple contributors. DNA of co-workers of the wearer and home partners were found in shirt samples. DNA of a male co-worker who had been on leave, and not in the office for 2 months, was found on the front of one T-shirt of a female participant.

A study investigated 'dragging' type situations where trouser ankles (48 samples) and armpits of shirts (16 samples) of known wearers after dragging events by known individuals were DNA profiled (van den Berge et al., 2016). DNA profiles contained 2 to 5 contributors and all indicated the presence of both the wearer and grabber.

An Australian study examined the transfer of DNA onto three 10×10 cm areas located on the front, back and shoulder of an individual's external clothing during a normal day's wearing (Ruan et al., 2018). The quantity of DNA in all three areas increased approximately eightfold, which usually corresponded with an increase in self DNA from the wearer on the front area of the shirt. However, the back area of the shirt was more likely to demonstrate mixtures of self and non-self DNA.

6.1.2 Location of DNA on Clothing

A decade ago it was found that DNA may transfer from item to item, and within parts of an item, in evidence bags during transport and handling after collection (Goray et al., 2012). Older, less-sensitive DNA profiling techniques were used in the experiment ('Profiler Plus' typing system using 28 amplification cycles). Further studies using the current more sensitive techniques are required to understand this aspect, especially in regards to DNA transfer from one part of a garment to another due to collection and handling processes, including transfer in an exhibit bag.

An interesting finding from the above study regarding blood (and thus a quantity of DNA) was an experiment that placed blood on the middle front *outside* of cotton underpants and allowed to dry for 24 hours (ibid.). Two underpants with blood from different people in the same deposit area were packaged together in a brown paper bag and handled as per transport to a laboratory. Different parts of each garment were tape lifted and had DNA analysis. DNA from the deposit area was found on each of the three non-deposit areas of the same garment in 100% of cases (front-outside, rear-outside and inside). DNA derived from the deposit was also found within non-deposit areas of the other garment. In some instances the major contributor to the mixture DNA profile corresponded to the other person. 'Surprisingly' the *inside* of the underwear contained the greatest proportion of transferred DNA compared to other areas. This was speculated due to the random placing and perhaps inversion of the underpants (ibid.).

The above study highlights the ready transfer of biological matter on exhibits when handling and transporting, and this book author recommends further study on trace DNA and caution when interpreting DNA on 'specific' areas of clothing. If DNA can transfer during transport and handling, it may also transfer during sampling and other forensic examinations – noting that gloves and other examination equipment such as forceps are efficient transfer vectors (Szkuta et al., 2015). The more an exhibit is examined the greater the risk of DNA transfer within that exhibit.

A case this author reviewed involved DNA profiles obtained from a shirt worn by a victim during an armed robbery by two offenders in a drinking/gambling venue.

CASE 6.1 'TOUCHING' A SHIRT

A security guard/gambling attendant was attacked in a gambling/drinking venue in the early morning just after the venue opened. The victim stated that one masked male 'offender 1' pushed the left-hand shoulder blade of his shirt with bare hands, and pushed him out of the bar area, also pushing him on his right upper arm. The other male masked 'offender 2' made forceful contact with the victim's left chest area with a bare left hand. The incident was captured on CCTV footage. Detectives (not crime scene officers) collected the shirt that morning at the scene, describing the victim as taking off his shirt and putting it in a brown paper exhibit bag.

Three areas of the long-sleeved shirt were swabbed in the local forensic science laboratory using wet and dry swabs – the rear left shoulder area, the right upper arm area and the left upper chest area (as indicated by police officers). Mixture DNA profiles were obtained of at least four contributors from the rear left shoulder and left upper chest, and of at least five contributors from the right upper arm which the laboratory report denoted as 'too complex: cannot be analyzed'. The two remaining DNA profiles were interpreted, conditioned assuming the victim as a contributor, using a probabilistic genotyping system and reference DNA samples from suspects A and B. A report was issued by the laboratory which considered as propositions H1 – the accused is a contributor and H2 the accused is not a contributor. (See Table 6.1.)

This author's review found that all of the corresponding components in the reference sample of suspect B and the victim were represented in both interpreted DNA mixture profiles, from the chest and from the rear left shoulder; however with different mixture proportions. All of the corresponding components in the reference sample of suspect A were represented in the rear left shoulder area. The propositions in the report were incomplete considering there were four contributors and alternatives were not explored (see Gill et al., 2018 for mixture propositions for sub-source level). The laboratory sampled the areas of the shirt they were directed to sample and did not use their own sampling and testing rationale or test 'control' samples on the garment.

The DNA results confused the interpretation by the justice system as 'activities' of touching were alleged by the prosecution, yet the DNA statistics seemingly contradicted the alleged

TABLE 6.1 DNA MIXTURE PROFILE RESULTS FROM SHIRT

Sample Description	DNA Profile Description	Person	Propositions/Interpretations	Statistical Weighting
Rear left shoulder tape lift of shirt	Mixed DNA profile – four contributors including the victim	Victim	Assumed contributor	
		Accused person A	H1. Accused A is the contributor H2. Accused A is not the contributor	The DNA evidence is 310 times more likely if accused A is a contributor.
		Accused person B	H1. Accused B is the contributor H2. Accused B is not the contributor.	The DNA evidence is 100 billion times more likely if accused B is a contributor.
Left upper chest tape lift of shirt	Mixed DNA profile – four contributors including the victim	Victim	Assumed contributor	
		Accused person A	H1. Accused A is the contributor H2. Accused A is not the contributor	The DNA evidence is 59 times more likely if accused A is a contributor.
		Accused person B	H1. Accused B is the contributor H2. Accused B is not the contributor.	The DNA evidence is 100 billion times more likely if accused B is a contributor.

scenario. Suspect B had a high likelihood ratio when assessing the mixture of DNA profiles from the rear shoulder and left upper chest, but was alleged to have touched only the left upper chest (offender 2). Suspect A was alleged to have touched the rear left shoulder and right upper arm (offender 1) but had low likelihood ratios (still greater than 1) for both the left chest and rear left shoulder.

During the pre-trial hearing the laboratory forensic scientist stated that it could not be determined how the DNA deposits arrived on the garment, whether by direct or indirect transfer. The expert also stated that DNA may have transferred in the exhibit bag when this was queried; however, this explanation was not noted in the laboratory report.

The case was resolved through plea bargains just before the trial.

It is important to consider how clothing items were removed from the body, packaged, handled and examined and if transfer principles were considered, such as avoiding folding parts of the garment against each other. *If* the prosecution scenario of Case 6.1 was true, then DNA must have either transferred on the garment after the offence or the hands of each of the offenders had non-self DNA of the co-accused. The victim removed his shirt and handed it to detectives (not crime scene officers) in the above case, so his hands and body may have been a vector for transfer. Similarly transfer may have occurred by folding in the exhibit bag, transport, handling and examination.

DNA does transfer – so detecting DNA on one *specific* area of a garment from examination in a laboratory cannot infer that the DNA had been deposited in that area during a *specific* incident (or activity).

A paper describing a template for Bayesian networks in forensic biology cases, when considering activity-level propositions, described a particular prosecution scenario and how to evaluate the evidence in light of alternative propositions from the prosecution and the defence using Bayesian reasoning and 'activity levels' (Taylor et al., 2018). The scenario provided in the paper was from a real case (see Conclusion of paper) but was not identified and it is unknown if it went to court or if the DNA was debated by the justice system.

CASE 6.2 DNA FROM UNDERPANTS AND ALLEGATIONS OF BITING

A 24-year-old girl (complainant C) who normally lived with her biological mother and father (F) stayed for a week at her older brother's (suspect D) house. A friend of the girl received a phone call from her stating that her brother (D) had bitten her on the vagina over her underwear. The friend picked up the complainant and went to the police where the underwear was seized. The brother was arrested and a reference DNA sample was taken. The underwear was examined by the local forensic science laboratory and the following was found:

1. Faecal staining was present on the inner and outer crotch of the underpants.
2. An RSID saliva test on the crotch of the underpants was positive.
3. A tape lift of the outer crotch of the underpants had a single DNA profile that matched the complainant. However, the quantification result revealed the presence of low levels of male DNA. The presence of the complainant's DNA in such high amounts meant that the male DNA was not able to be profiled using autosomal profiling systems.
4. Y-STR profiling of the outer crotch tape lift extract yielded a single-source Y-STR profile that matched the brother's Y-STR reference profile.
5. A tape lift of the outer front of the underwear yielded a single-source routine DNA profile that matched the girl. The quantification result revealed no male DNA and no further work was carried out on the item.

It was assumed for the results calculation that the underwear was appropriately seized, packaged and stored – but case studies in this book show this is not always the case.

The competing propositions to describe activity levels of the prosecution and defence scenarios were:

Hp: D has bitten C on the vagina, over her underwear.
Hd: C has been staying at D's home, but no biting occurred.

The case study notes that the results of 'similar' samples were grouped together because DNA does transfer. The findings of the two tape lifts of the outer surface of the underwear were considered as being one large tape lift of the entire outer front and crotch of the underwear and that low levels of male DNA were present. It is noted by this author that the definition of 'similar' may be debated – as Case 6.1 demonstrates – and transfer of DNA is described from areas of garments in exhibit bags (Goray et al., 2012).

There was no study that directly examined saliva from biting clothes that could be used in the case evaluation, so results from other studies were used. Another limitation of the evaluation was that only the false positive of faeces to the RSID saliva test was considered (as well as the level of background saliva). The supplementary material of the paper showed that the probability of obtaining a positive RSID saliva test in the presence of faecal material was assigned as 0.67 for 'yes' and 0.34 for 'no' based on one study (Casey et al., 2010). That study tested the faeces of four infants and found that three gave a positive result for the RSID saliva test. Probabilities were deduced as (3 + 1)/(4 + 2)~0.67 for the node state 'positive' and (1 + 1)/(4 + 2)~0.33 for the node state 'negative'. The low numbers in the referred study of just four participants (and infants at that) demonstrate the problem with relevant published studies as outlined previously (Gosch and Courts, 2019; van Oorschot et al., 2019).

An Australian study on intimate samples from females noted that vaginal swabs, vulva swabs and the crotch area of underpants were excluded from their own laboratory saliva screening tests (including RSID saliva) as they could give positive results in the absence of saliva (Sari et al., 2019). Thus in some laboratories the underpants of females would not be examined at all for saliva. This calls into question the rationale for testing for saliva on the underpants in Case 6.2.

The likelihood ratio for the probability of the results given the two alternative propositions of how the DNA (profile and amounts) was deposited, given alternative propositions of biting the underpants or innocent transfer by co-habitation, was about 0.19. That is, the results are about five times more probable given the defence proposition than the prosecution proposition (Taylor et al., 2018).

The authors of the article highlight that a typical 'sub-source' level report would state there was a positive result for human saliva on the crotch area of the underpants and a Y-STR profile that matched the defendant.

> Even if the report went on to detail the possible causes of false positives (or coincidental matches) for the findings, if it did not place those in the context of activities one could see how the results would appear to a lay jury to strongly support the prosecution version of events over the defense version … the presence of faecal staining on the underwear lends very little support to either one of the propositions over the other. Additionally the activity of biting would lead to an expected high level of DNA transfer, whereas only a low level of matching DNA was detected and this could be from the father of the complainant … the result is contrary to how the lay jury is likely to have interpreted the results without any guidance from the scientist. Such a difference can have a major impact on the outcome of a trial. (ibid.)

This current book author notes that the relevance of findings, even if sub-source, should be communicated in clear language including the assumptions and limitations of the results. The examination of the underwear for saliva when there is known reactivity of the test used with both faeces and vaginal secretions, and the searching for male DNA that could be present due to co-habitation of family members, should raise red flags to the reporting scientist in the first instance. It should not have to resort to a trial situation considering activity-level propositions to realize the limitations of the case to the reporting scientist.

Conclusions about the source of DNA should not be conflated with conclusions as to how it arrived there and more research is required to strengthen approaches to activity assessment. Although potentially useful to investigators or fact-finders should the scenario and protocol at issue in the instant case be identical, the overall variability in these types of studies makes it currently problematic to render broad conclusions about expected previous activity or item use based on residual detected amounts of DNA or DNA profiles (Burrill et al., 2018).

Bayesian reasoning is increasingly being discussed in the forensic DNA literature, especially for 'activity level assessments' which may be important for clothing exhibits and personal items (i.e., Gill et al., 2020; Hicks et al., 2022; Taylor et al., 2018; van Oorschot et al., 2019). The advantages of Bayesian networks allow the expert to model all relevant parameters and their dependencies. The graphical visualization of complex Bayesian formulations enhances the transparency of the expert's reasoning. The model also explicitly requires experts to assign probabilities to all modelled parameters, thereby avoiding implicit assumptions. However, even if this reasoning is appropriate current knowledge regarding specific scenarios may render any assessment speculative or irrelevant, as demonstrated by Cases 6.1 and 6.2.

6.1.3 'Touching' a Garment by a Person of Interest

6.1.3.1 Clothing in General

The transfer of trace DNA to clothing has been investigated recently in literature studies using a number of 'typical' scenarios encountered in case work. Simulations focused on obtaining DNA from the perpetrator in attack, aggression and grabbing and dragging scenarios, have analyzed DNA from the clothing worn and frequently detected DNA that could not be attributed to the wearer or perpetrator on external areas sampled. Whether a result of direct or indirect contact with an individual, or through exposure to the surrounding environment, these studies demonstrate that background DNA – relative to the targeting of an individual (wearer or other) – is typically collected during sampling and observed in the resultant DNA profile (Szkuta et al., 2019). Spending time in a shared space (such as home or work) may result in the transfer of DNA from others who occupy that space, without these persons contacting each other.

A study was conducted by four laboratories as it was stated that there was limited understanding of how, and from where, background DNA is derived on clothing (Szkuta et al., 2020). Research on the transfer, persistence, prevalence and recovery (TPPR) of DNA traces from upper garments was conducted and samples collected from several areas of two garments, each worn on separate working or non-working days and individually owned by four individuals from each of the four laboratories, and processed from DNA extraction through to DNA profiling. Questionnaires documented activities relating to the garment prior to and during wearing, and reference profiles were obtained from the wearer and their close associates. Variation in the DNA quantity, composition of the profiles and inclusion/exclusion of the wearer and their close associates was observed among the collaborating laboratories, participants, garments worn on different occasions and garment areas sampled (ibid.).

The above study is very interesting as it emphasizes the multiple variables that need to be considered when examining clothing as exhibits in a criminal case. The amounts of DNA deposited varied from 0.002 ng to 68 ng from the garments, a difference of over 10,000 fold. Two-person mixtures were more frequently observed in Lab 1 (48.4% of profiles generated by this lab) and Lab 2 (39.1%), while three-person mixtures were more frequent in Lab 3 (43.8%) and Lab 4 (42.2%). Further, there was a consistent trend across the four laboratories of an increased minimum number of contributors in profiles generated from the external back, breast and inner forearm (typically two- to four-person mixtures), while samples obtained from the external armpit and internal collar (Labs 1, 2 and 4 only) resulted in fewer contributors (typically single-source or two-person mixtures). The study postulated that the extent to which DNA-containing material resides on public and private items explains the increased minimum number of contributors in profiles obtained from external areas of upper garments within Labs 1 and 2, such as the back and inner forearm, given the frequency by which these areas come into contact with various surfaces. Internal areas such as the collar and cuff, and those that are generally shielded (e.g., armpit), typically avoid contact with the surrounding environment and thus the transfer of non-self DNA to these areas is minimized (ibid.). Note a comparison can be made with Case 6.1 and the five-person mixture from the upper arm area of the shirt in the case with the study findings.

The study also found 'surprising' the extent to which partners and children were observed as contributors to profiles generated from clothing worn during working days, especially since garments were worn at work for a further 8–11 hours after being present in the shared home. This suggested that DNA from those inhabiting with a person may persist, and be detected, on their clothing over a long period of time, even after many activities are performed and contacts made with the clothing item (ibid.).

6.1.3.2 Underwear and Purported Sexual Assaults

Clothing, particularly underwear, from the complainant is often obtained as exhibits in sexual offence cases, in addition to medical exhibits (see Chapter 5 for medical exhibits). Underpants and lower garments such as jeans and shorts may be examined for the presence of semen and DNA from a potential offender. This may be due to direct ejaculation onto the body or garment as well as drainage from internal ejaculation (this may also refer to male/male rape). Upper garments from a complainant may be examined if there has been an oral sexual assault. Screening tests such as the acid phosphatase (AP) test are used to detect areas of potential semen staining and these are then sampled and analyzed for the presence of sperm and DNA.

Understanding of trace DNA on the undergarments of a complainant in sexual offence cases is complicated by the fact that just 'DNA' cannot infer sperm/semen. This means that the DNA could have been deposited by innocent means such as through social or domestic interactions, or with no interaction at all. Clothing may have direct and indirect deposits of trace DNA from many times since the last wash. Clothing may be removed by the suspect and end up on the bed, furniture or even the floor.

Sampling clothing for potential deposits of trace DNA from an offender during a sexual assault is also complicated as there is no 'screening' test for such. The examiner is reliant on information provided by the complainant, previous knowledge or the (limited) literature. Sampling too large an area (such as tape lifting the entire pair of underpants and combining the tape lifts) may result in un-interpretable mixed DNA profiles. Sampling too small or the wrong area may result in too little DNA to be analyzed.

A study on the background levels of DNA that underwear may pick up during a typical day of wearing was undertaken (Murphy et al., 2020). The inside front of clean, but not new, pairs of underpants worn for a day by a female – who was asked to abstain from sexual intercourse for 48 hours prior to taking part in the study – was analyzed for male DNA. There were 83 samples from a total of 103 that showed the presence of male DNA in varying quantities and 24 contained sufficient DNA to proceed to profiling. Of those 24 profiled samples, 12 samples detected the Y chromosome and varying numbers of minor alleles (not from the female wearer), 5 gave a complete Y-STR profile and 18 gave incomplete Y-STR profiles. Each of the underpants samples that gave a complete Y-STR profile was from a female who cohabited with a male partner and the Y-STR profile matched that of the partner where a reference profile was available (four out of five). The authors concluded that in the absence of sexual contact, a Y-STR profile, if present, is more likely to be obtained from a cohabitating male rather than from a non-cohabitating male in social contact with a female.

One study examining trace DNA transfer within specific scenarios that can occur during sexual assaults was performed using a 'dummy' wearing various items of undergarments (Ramos et al., 2020). Note that this study did not examine the movements of a human during the mock scenarios (the 'victim'), nor their own DNA that would be present and complicate the interpretation of DNA profiles obtained.

The study (ibid.) recorded how much DNA was transferred by volunteer 'offenders' during six types of mock sexual assault scenarios to underwear. These were removing underpants, removing a brassiere, digital penetration of the vagina from the front, grabbing breasts over the top of the brassiere, digital penetration of the vagina from the rear, and grabbing the breasts under the brassiere from the rear (ibid.). The study demonstrated that some areas corresponded to areas typically sampled during casework and others less so – thus the study provides information to help improve sample targeting. The study also observed several instances of a colleague or cohabitating partner of mock offenders being detected within a DNA sample – demonstrating indirect transfer using hands as an intermediate vector. Likelihood ratios of greater than 10,000 were signified as a contribution of DNA by an individual *(author note – laboratories reviewed by this author denote likelihood ratios less than 100 as being meaningful as to contribution, see Chapter 2)*. One simulation had a two-person DNA mixture from the top cups of a brassiere where the 'offender' was not present, the domestic home partner of the offender accounted for the minor contributor (33%), and an unknown person accounted for the major contribution (67%).

The study (ibid.) also determined that the amount of DNA that is deposited on underwear is highly variable and depends largely on the individual undertaking the action and then secondarily the activity they are undertaking. The amount deposited varied from 0.00018 ng to 12.3656 ng, a difference of over 10,000 fold. Even when the activities were expected to only produce contact on one side of the underwear (e.g., the inner surface for the digital penetration scenarios), DNA from the mock offender could still be detected on the other side (although usually at lower amounts). This was explained because this area was contacted during the activities, or it may be due to DNA being transferred from location to location on the item when or after it is packaged in a brown paper bag (ibid.). Also see Case 6.1 for potential alternatives.

Note that the above study examines trace DNA transfer, where the DNA cannot be related to a specific biological fluid or incriminating 'sexual activity'. An older study examined the reverse situation of the transfer of female trace DNA to the underpants of a male (Jones et al., 2016). A body fluid forum group in the United Kingdom and Ireland performed research to assist in cases where semen is absent on vaginal swabs but female DNA is present on penile swabs or male underwear, and the issue to be addressed is whether or not sexual intercourse occurred (see Chapter 5 for review of the penile swab experiment). Note that this study used less-sensitive techniques than that of today.

The study above (ibid.) simulated urination by the male participant after close social contact (not intimate) for 5 minutes of a male and a female and the male underpants were then analyzed for DNA. There were five occasions out of 30 simulations where female DNA was obtained from the waistband of the underpants and one where there was female DNA in an outside front panel of the underpants, but only one single allele matching the female participant.

The murder of Meredith Kercher and the accepted miscarriage of justice (Case 3.2) gave impetus to the study of tertiary transfer from surfaces similar to a door knob to surfaces similar to a bra clasp, via gloves as a vector (Fonnelop et al., 2015a). It is always important to remember the collection and handling practices at crime scenes as well as in the laboratory that may be inadvertent vectors for DNA transmission.

There needs to be further research into undergarments and DNA obtained from them during non-intimate activity with the advent of more sensitive technology, and using more realistic scenarios – especially if data are to be used in 'activity' level evaluations. This may be difficult to achieve due to ethical considerations.

6.1.3.3 Clothing over Skin

There had been no studies on the transfer of 'offender DNA' deposited on the skin of a victim onto clothing worn over the assaulted area until a study in 2018 (Bowman et al., 2018). This study suggested that sampling from clothing worn over the assaulted area may be an additional or better avenue for the recovery of offender DNA post-assault where there has been significant time between assault and sampling.

6.1.3.4 Clothing in Motor Vehicles

The interior of motor vehicles may be targeted for trace DNA. Cars may be used in many criminal activities including personal assault (sexual and/or violent assault) as well as using them as get-away vehicles. In the case of personal assault it may be important to recover biological matter that can be related to semen or blood. However, many cases also require the analysis of trace DNA to determine any indication of a person's DNA. Chapter 4 describes the importance of understanding the level of pre-existing DNA in the environment of the car.

A study investigating case sampling strategies when investigating DNA transfer to *the clothing* within vehicles found that in many instances DNA from a regular driver of a vehicle could be recovered from samples collected from the external back of the upper garment, and from the seat of the trouser pants, worn by an incidental driver (30 minutes) of that vehicle soon after departing the vehicle (De Wolff et al., 2019). This highlights a potential avenue of indirect transfer that may be relevant during investigations of particular types of offences.

The relevance of fibre papers regarding secondary and higher transfer should not be ignored, and this book author recommends a holistic understanding of trace evidence transfer (see Chapters 1 and 2). The transfer of fibres whilst sitting in movie cinema seats and other seating has already been documented in the literature many years ago.

It is also important that consideration be given to techniques such as 1:1 taping that have been used for fibres in cases of clothing on the bodies of the deceased. There is no reason why this technique cannot be used to trace DNA.

6.1.4 Clothing on Flooring

Clothing may be placed on floors when removed from the body, whether during a criminal act or during everyday wear. Flooring may also be contacted by garments when worn if a person fell to the floor or was forced down to the floor.

Shoes, socks and bare feet regularly contact the flooring of a home providing an avenue for the transfer of DNA. Further, bags and other personal items such as tissues and other items laden with DNA may drop or be placed on the floor.

Flooring surfaces within homes contain large quantities of DNA from occupants who did, and did not, use the room (Reither et al., 2021) and also from rooms that had been previously cleaned (Reither et al., 2019). DNA may be obtained from floors from visitors who had not been to the house recently or had moved out months prior.

A study investigated the transfer of DNA to and from flooring in a residential environment (Reither et al., 2022). Two scenarios were investigated – clothing placed on the floor or worn when the person lay on their back on the floor and rolled, with the main bedroom targeted. DNA transfer was observed in both scenarios (ibid.). A broad range of DNA quantities were recovered from areas from the same garment, garments worn by the same person and garments worn by different wearers. DNA was recovered only in areas of contact with the floor (ibid.).

6.1.5 Washing of Clothing

The persistence of DNA on clothing after washing/cleaning has been investigated in recent years, perhaps due to the increased sensitivity of DNA-detection methods.

Initial studies investigated the persistence and transfer of DNA from body fluids, such as semen, blood and saliva. Early studies showed that spermatozoa may be detected on items previously washed with a semen-stained garment (Kafarowski et al., 1996).

DNA profiles from semen stains were found to still be retrievable after washing and also retrievable from other co-washed clothing as well as the washing machine drum (Noel et al., 2016). This study found that of 168 samples taken from underwear (n = 24) of children that had been regularly worn and washed with the rest of the family's laundry, 52% yielded interpretable mixtures of DNA corresponding to multiple family members (including fathers, brothers, sisters and mothers) post-washing and drying. DNA corresponding to the mother was detected in 51% of samples, contributing 5% to 90% of the total genetic mixture.

An interesting case study showing the persistence of spermatozoa on clothing is from Japan (Ogawa et al., 2018).

CASE 6.3 DRY-CLEANED SKIRT

A dry-cleaned skirt in a sexual assault case was examined. The victim was a female high school student who had been sexually assaulted outside a small wooden warehouse at night. She had worn the skirt for more than a year after the assault and it was not possible to confirm how many times the skirt had been dry-cleaned – the skirt had also been subject to stain removal treatment. An examination on an external genital swab after the victim had showered was performed which identified a suspect but during the trial the defendant's lawyers raised doubts. The court had located a police photograph of the skirt which depicted white staining on the interior front. The court ordered an examination of the skirt for semen and DNA. Both alternative light source screening and screening tests for semen (AP) were negative. Samples of fibres were cut from areas approximated from the photograph and spermatozoa were located (about 10 to 20 with partial tails).

Three different multiplex systems were used on the sperm cell fraction (including Y-filer) and DNA profile obtained that were consistent with the reference sample from the suspect.

Informative DNA profiles have been obtained from blood-stained washed cloths and from co-washed clean cloths (Kulstein and Wiegand, 2018).

The recognition of traces of material from washed clothing is the first consideration for any subsequent interpretation of DNA profiles. If a garment is not sampled because there is no obvious stain or deposit then there is nothing to interpret. Alternate light sources including infra-red light are useful for dark fabrics where blood stains may not be visible, and on washed clothing with original deposits of blood. Experiments have shown alternate light sources useful in detecting blood deposits on fabrics up to the fourth or fifth wash depending on the fabric composition and construction (James, 2021).

Studies have also focused on the transfer and persistence of skin-derived biological material on washed clothing. Although some studies concluded that DNA persistence and transfer in a washing machine was unlikely further studies found different results, which may be due to the increased sensitivity of DNA-testing kits (van Oorschot et al., 2019).

DNA transfer may occur during the whole process of washing and drying clothes and not just within the machine itself. Garments are placed in laundry hampers prior to, and after, washing and there are various ways of drying clothes such as on racks or lines, all potential vectors of DNA transfer.

6.1.6 Fabric Gloves

Gloves are a special item of clothing as they are worn on hands to protect them from the environment, such as the cold (fleece or knitted gloves), heat (oven mitts), rough surfaces, soil and trees (gardening gloves), dishwashing and cleaning (rubber gloves), for driving cars and riding horses (leather or fine synthetic fabric gloves) and for working in the forensic laboratory and at crime scenes (disposable nitrile gloves). Hands are accepted as a continual means of contacting the wearer's own body and face, and other items, so the accumulation of DNA on gloves (not disposable or cleaning gloves) is to be expected.

Laboratory gloves have been much discussed in the forensic literature due to their constant use in forensic exhibit collection, handling and examination. These gloves need to be disposed between item-to-item examinations, and often between separate parts of an item, due to their known propensity as an efficient transfer vector. Double gloving has also been recommended. A lack of strict observation of this requirement may result in flawed examinations (Fonnelop et al., 2015a; Gil, 2016).

Gloves worn in everyday activities other than in forensic exhibit collection may end up as exhibits in a criminal case. Fabric gloves, especially if infrequently cleaned, may be a reservoir of DNA from many times and many contacts. Gloves may be worn in burglary cases to avoid leaving fingerprints at the scene or for protection. An interesting German study examined 'working gloves' to simulate a burglary using a screwdriver as a break-in tool, as a result of a specific criminal case (Otten et al., 2019).

CASE 6.4 GLOVE FOUND AT SCENE OF BURGLARY (OTTEN ET AL., 2019)

Investigators found a screwdriver and a working glove in the garden behind the house after a burglary on a private residence. Both the screwdriver and the glove were subject to DNA analysis and the results were compared to a suspect DNA profile. The screwdriver had a DNA profile matching that of the DNA profile of a suspect, and the glove had a mixture DNA profile containing fragments (alleles) also shared by the suspect. Likelihood ratios at the sub-source level were obtained that showed extremely high support in favour of observing the DNA profile given that the suspect was a contributor, in comparison to observing the DNA profile given that he was not a contributor, which was a likelihood ratio greater than one million (ibid).

During the court hearing the defence submitted that the accused was the original owner of the glove and a handy man, had recently lost the gloves during a house-moving job, did not break into the house and the screwdriver was not his. It was suggested that an unknown person used the glove to handle the screwdriver, forcing open a window in the house. This changed the value of the evidence from sub-source level to the activity level as the source of DNA was not disputed. Thus the prosecution hypothesis was 'the suspect handled the screwdriver wearing gloves' and the defence hypothesis 'an unknown person (the true perpetrator) wore the suspect's glove while handling the screwdriver'. Due to a lack of data on the specific scenario, the expert was not able to assist the court in assessing the likelihood ratio at the activity level (ibid.).

Studies were performed by the authors to assess the amount of secondary DNA transfer of an innocent person's DNA to a crime scene via working gloves taking into account shedder status (Otten et al., 2019). Note the previous discussions in Chapter 4 regarding the concept and most likely 'continuum' of shedder status and the debate about what exactly is in 'trace DNA'.

The authors assessed the shedder status of participants in the study by following a Norwegian protocol describing 'good' and 'bad' shedders (Fonnelop et al., 2017). A total of 40 participants were paired according to shedder status, with 12 'good' and 25 'bad' and three not determinable. Even so, values were excluded from the study due to four participants depositing 'incomparably high' amounts of DNA. 'Shedding ability' varied from other studies in the forensic literature, such as age, time since last hand wash and mixture profiles produced.

One simulation experiment was a house move and the other experiment simulated a burglary, with both participants wearing the same pair of working gloves. The gloves consisted of 100% nylon fabric with an additional latex coat on the palm and fingers, typical of working gloves encountered in the country. Primary transfer described the transfer of trace DNA from the palms of participant 1 (P1) and participant 2 (P2) both onto the outside and the inside of the gloves with P2 being the last handler. Data suggested that the shedder status determined the major contributor to the DNA profile deposited via primary transfer, regardless of the order in which the participants handled the object, contrary to some other studies such as the last handler being the major contributor or duration of contact.

The secondary transfer described the transfer of trace DNA from P1 or P2's palms onto the handle of the screwdriver with the glove (especially the outside) acting as a vector and P2 being the only one to handle the last object. Results suggested that it is possible to *not* observe DNA from the person handling a tool while observing DNA from a person who never handled the tool. DNA from an uninvolved person (P1) was transferred onto the object via secondary transfer by conventional working gloves. However, a significantly higher amount of profiles with poor quality (low peak heights and allele and locus drop-outs) were found on the screwdrivers compared to the outside and inside of the gloves (ibid.).

The main purpose of the study was to test whether or not P1's DNA could be detected on a screwdriver even though this person was never in contact with it. Findings clearly showed this is possible with a probability of 0.31 (6 out of 19) irrespective of shedder status. If P1 was a good shedder this probability increased to

0.6. Six experiments out of the 19 in total in which the deposited DNA on the break-in tool 'matched' the DNA profile of a person who was never in contact with the tool (P1), had a likelihood ratio greater than one million. The authors stated that before using these findings in court much larger sample sizes and more in-depth experiments are required (ibid.). Nevertheless, their findings show that the initial defence hypothesis was not disproved and indeed was feasible.

6.2 Knives

6.2.1 Direct Transfer of DNA to Knives

The study of DNA transfer to knives has become a topic of particular interest due to the use of knives as a weapon in violent crime, particularly in the United Kingdom, many countries in Europe and Australia where possession of firearms are subject to strict law enforcement.

There is still little research which examines the presence of background DNA on items before a criminal event takes place. This is necessary to consider with knives and pre-existing DNA before use by a potential offender.

It has been described that DNA is readily transferred to a knife handle by hands during a stabbing action and DNA existing on the handled knife is readily picked up during the action and transferred to a subsequently handled object. An Australian study aimed to determine which of the multiple last known handlers of a knife could be detectable by DNA analysis (Buckingham et al., 2016). Sets of four individuals each consecutively handled a knife by simulating a stabbing action and placed their hands on glass plates and the DNA collected analyzed. Each of six different situations had the DNA detected of all four individuals. The last handler was not always the major contributor. Proportional contributions to DNA profiles retrieved from knife handles varied depending on the individual who touched the knife handle.

A further study from the same authors (Buckingham et al., 2017) repeated the same type of experiments but the particular hand prints were on cotton plates instead of glass plates. Less DNA was collected from the cotton which was said to be due to less efficient recovery of DNA. The DNA profiles of the later handlers of a knife were more prominent than earlier handlers in samples retrieved from subsequently handled objects, and the quantity of foreign DNA picked up by a hand and deposited on subsequently touched objects diminished as more DNA-free objects were handled soon after each other (ibid.).

A Swiss study examined stabbing simulations with a stainless steel knife (Samie et al., 2020). This study described that the DNA shedding ability of an individual should be characterized as a distribution of the quantity of DNA present on hands or transferred on surfaces, and that individuals do not have a fixed shedder status such as 'good' or 'bad', in contrast to the German study above for gloves (Case 6.4). The Swiss study (Samie et al., 2020) dealt with two situations regarding a knife handle. The first was a primary transfer from a hand to a knife handle, and the second was a secondary transfer from a person of interest (POI) to the hand of an intermediate person who then took the knife handle (ibid.).

Examining primary transfer each participant rubbed their hands together, took a knife handle with their usual hand, and then stabbed a ballistic soap. The duration of the contact, the type of contact and the force of stabbing were not specified in order to simulate conditions as close as possible to casework, and the ballistic soap was used to mimic a human body. The entire surface of the knife handle and the inside surface of the other hand were swabbed to collect DNA. A large variation in the quantity of DNA collected from the participant's hand and the quantity of DNA transferred to the knife was observed between participants (ibid.). Further, a large variation was also observed for each participant, such as from 0 to 5 ng from the knife handle, and from 0 to 11 ng from the hand from one participant, depending on the experiment. This accords with variations in what is known as 'shedder status' for one person, at different times and for different experiments (see Chapter 4). Note also the few numbers of participants (n = 6) in the Swiss study.

An important finding was that there was *no obvious relationship between the quantity of DNA of the person of interest recovered on the hand and the quantity of transferred DNA* (ibid.). For example, participant 3 had a large quantity of DNA on their hand compared to other participants yet transferred a very small quantity of DNA to the knife handle through primary transfer (ibid.).

Participant 1 had small quantities of DNA on his hand yet transferred a large part of that DNA to the knife handle. This paper stated that *for primary transfer there is no fixed transfer proportion for participants* (current author italics). The finding is relevant to all surfaces purported to be contacted by hands including other implements such as firearms.

Findings also included variations between donors with regard to the non-self DNA present on the hands and transferred to the handle – for example, 40% of the total quantity of DNA on the handles used by participant 1 came from someone else (ibid.). The authors question the use of the terms 'good shedder' or 'bad shedder' and instead propose an analysis based on the mean and standard deviation of a person's shedding characteristics. Again (see Chapter 4), this book author questions the ethics of forcing a suspect (or indeed a complainant) into a study on their shedding ability when they have already donated DNA under legislation.

An Australian study (Szkuta et al., 2017) observed between 0.1 and 85.5 ng of DNA on hands (n = 70). Another study (McColl et al., 2017) observed higher quantities between 0 and 585 ng with a larger sample size (n = 120). The Swiss study (Samie et al., 2020) observed a total quantity of DNA on the hands between 0 and 21ng made mostly of the donor's own DNA, but the percentage of non-self DNA varied substantially.

Other researchers have reported higher quantities of DNA directly transferred to knife handles. Two Australian studies reported between 3 and 10 ng of total DNA on knife handles (Meakin et al., 2017) or between 1 and 10 ng of total DNA (Butcher et al., 2019).

The Swiss study (Samie et al., 2020) stated that the transfer probability (TP) varies between participants and depends on the type of transfer (primary versus secondary, for example). This means that it is not feasible to resort to a quantity of DNA on the one hand to infer the shedder status, and then assess what will be transferred to a surface. It is recommended that in order to evaluate a given quantity of DNA in light of activity-level scenarios, it is necessary to measure empirically the appropriate underpinning distribution dependant on the donor, the substrate and the transfer mechanism (Samie et al., 2020).

It is important to note that knives, although mass produced, come in varying shapes, sizes and compositions. Some have plastic or rubber handles and others metal. Blades may be stainless steel with or without serrated edges or perhaps double-edged. Handles may be relatively large for a full grip by a hand, or small due to the convenience (e.g., a folding 'Swiss Army' knife). There needs to be more detail in the literature regarding this variability to enable a sufficient comparison when evaluating data.

6.2.2 Indirect Transfer of DNA to Knives

An English study investigated whether secondary transferred DNA could be detected on knives that were regularly used by one person, and if so how long that might persist (Meakin et al., 2015). Each of the four volunteers handled a knife for 1 minute twice a day for two consecutive days. Then once a day they then shook hands with another volunteer for 10 seconds, and then immediately repeatedly stabbed one of their regularly used knives into a foam apparatus for 1 minute. Mini-tapes were used to collect the DNA from the handles. Mixed DNA profiles of at least three people were recovered from the knife handles an hour after the handshaking and stabbing sessions. These mixtures could be attributed to DNA from the specific regular user and hand-shaker for each set of knives and to DNA from unknown sources. DNA profiles from the hand-shakers were observed as minor profiles. The fourth pairing only complete single-source DNA profiles from the user were recovered. Where indirectly transferred DNA was observed, this persisted for at least a week (ibid.).

A further analysis of the above study was undertaken by the same authors (Meakin et al., 2017). They again showed that regular handling of knives could give detectable levels of regular user DNA that persisted for at least a week, varied between individuals and did not always result in complete DNA profiles. This is a reflection of our understanding of trace DNA transfer, persistence, prevalence and recovery. The knives were steak knives with plastic handles. There appeared to be more variation between individuals in depositing their own DNA than between samples from the same individual. Person A was observed as a minor contributor (~10%) to the mixture DNA profiles retrieved from the knives of three pairings, but not on knives of the fourth pairing (ibid). Recovery of non-donor DNA from handled items was not an unexpected observation and the authors considered a common feature of 'touch samples' (ibid.).

One United States experiment maximized the potential of transfer of DNA from person A to person B, and then from person B to the handle of a knife (Cale et al., 2016). Persons A and B washed and dried their hands, then wore a glove for 1.5 hours prior to shaking hands vigorously for 2 minutes, then B handled a knife handle for 2 minutes. This study found that detectable DNA of person A transferred to the knife on 85% of occasions. In five samples the DNA profile resulting from secondary transfer was either the only contributor or the major contributor identified despite never having come into contact with the knife, which has the potential to falsely link an individual to an item of evidence. The study emphasized the risk of assuming that DNA recovered from an object results from direct contact (ibid.).

A Swiss study described previously regarding the primary transfer (Samie et al., 2020) also examined secondary transfer by stabbing simulations with a knife. Like primary transfer, secondary transfer also had a large variation between participants. Two participants were chosen based on the first experiments as 'the best' or 'the worst' DNA donors. Both participants were first asked to shake hands and then to stab the ballistic soap with a knife. The contact did not exceed more than a few seconds. A distinct difference was observed from the two participants between the amounts of DNA recovered on the knife handle. The quantity of DNA is generally reduced when moving from hand (the quantity collected when sampling directly from a hand) to primary transfer (the quantity collected from a surface handled by a hand) and, subsequently, to secondary transfer.

A paper in 2018 noted that there is little consistency in the literature on when and how indirect transfer will occur, how much DNA may be transferred or how many interaction events such transfer will persist when studying DNA transfer to knife handles (Burrill et al., 2018), which is still true today. Studies investigating DNA recovery under certain deposition or transfer scenarios have demonstrated such wide variability that conclusion statements inevitably advise caution in the interpretation of these results in casework.

6.3 Firearms

6.3.1 DNA on Firearms

A lack of knowledge regarding the composition of trace DNA profiles on firearms was confirmed by one study using DNA-TrAC, a previously devised and freely accessible and searchable database on DNA transfer studies (Gosch and Courts, 2019). Experiments were performed to provide further information (Gosch et al., 2020). Trace DNA was recovered from various surfaces of two types of firearms, handled in four scenarios that might be formulated as alternatives in gun crime. The firearm (pistols and revolvers) had background DNA deposited over 4 days by the mock owner such as transporting the firearm in a bag and placing it at home amongst possessions.

The four different activities then undertaken with the firearms included: the mock owner performed a shooting and disposed of the weapon immediately; a second individual performed the same shooting scenario as the mock owner; a second individual imitated a longer and more intensive contact with the firearm including carrying it around in his waistband; a second individual performed the same shooting scenario as the mock owner, then wiped the firearm using a dry cotton towel for 15 seconds simulating fingerprint removal (ibid.).

Amounts of deposited DNA demonstrated intra- and inter-individual differences (ibid.). Observed differences between handling conditions, firearms and surface types, handling individuals and individual deposits emphasized the complexity of trace DNA profile composition expected to be recovered from firearms after realistic handling scenarios. Two cases of an indirectly transferred major contributor (> 90% profile contribution) were observed, which were identified as the shooter's girlfriend and the previous handler of the wiping towel, highlighting the relevance of previous contacts and activities prior to direct contact with the firearm (ibid.).

The importance of prevalent or background DNA was emphasized in a case reviewed by this author.

CASE 6.5 DNA ON FIREARM FROM GIRLFRIEND OF ACCUSED

A man pleaded not guilty to a charge of attempted murder for shooting at his uncle's partner and her friend whilst they were in a car. The woman said that she was in the driveway of her home reversing her car, with a friend in the passenger front seat. She claimed her partner's nephew in another car pointed a gun and fired numerous bullets into her back windscreen. Both women were later taken to hospital with minor injuries.

The home of the girlfriend of the suspect was raided by detectives and the girlfriend was arrested. During a search of her house a gun was found in a bedside cabinet adjacent to the bed in the main bedroom. The gun was in a bag in the drawer. The bag was not examined for DNA. It was uncertain what if any mitigation of DNA transfer was performed at the house or if the detectives even wore gloves.

A DNA analysis in the forensic science laboratory was performed on swabs from the cocking handle of the gun. A review by this author revealed the DNA profile from the gun was a partial and mixed DNA profile of at least three contributors, with low-level components. The laboratory report stated it was estimated to be 820 million times in favour of Hp, that the male accused is a contributor; that is, extremely strong support that the accused is a contributor to the DNA.

A reference DNA sample of the girlfriend of the accused was subsequently requested by defence before trial, due to the location of the gun and the possible prevalence of background DNA. A likelihood ratio was obtained against this reference DNA profile that was estimated to be 1.1 billion in favour of Hp, that the girlfriend is a contributor to the DNA. A summary of the results is presented in Table 6.2.

This author determined that if both the accused and the girlfriend were considered to have contributed to the mixed DNA profile, there must have been at least four contributors – which at the time could not be calculated by the forensic laboratory. The laboratory maintained there were three contributors. The DNA evidence was not admitted at trial.

The fact that the girlfriend of the accused produced a higher likelihood ratio for contribution for observing the DNA on the cocking handle of the gun than the accused illustrates the importance of not conflating a DNA profile or likelihood ratio with an activity, such as shooting a gun.

TABLE 6.2 DNA MIXTURE PROFILE RESULTS FROM COCKING HANDLE OF A GUN

1.Sample Description 2. DNA Profile Type	Person	Hypotheses/Interpretations	Statistical Weighting	Verbal Equivalent
1. The cocking handle was sampled for any traces of DNA that might be present.	Accused	Hp: Accused is a contributor to the DNA. Hd: Accused is not a contributor to the DNA.	820 million times in favour of Hp.	Extremely strong support for the hypothesis that the accused is a contributor (Hp)
2. Mixed partial DNA profile – three contributors.	Girlfriend	Hp: Girlfriend is a contributor to the DNA. Hd: Girlfriend is not a contributor to the DNA.	1.1 billion times in favour of Hp.	Extremely strong support for the hypothesis that the girlfriend is a contributor (Hp)

The accused was found not guilty at trial. After the verdict the police prosecuted the (now-ex) girlfriend for possession of a gun and she was found guilty.

The above case highlights the importance of not conflating the presence of a DNA profile (sub-source) on a gun with the *activity* of shooting that gun. It may also be of interest to perform studies regarding intra-exhibit transfer of DNA on guns while handling and in transport and examination – the relevance of trace DNA on a 'cocking handle' may not be what was interpreted by the prosecution. The bag containing the gun was also not examined for DNA, and it has long been known that DNA can transfer from bags to exhibits within them (Goray et al., 2012).

A very interesting USA case that scrutinized laboratory staff proficiency and who could be 'excluded' from a DNA mixture focused on DNA analysis on a firearm (*USA v. Kenneth Davon Lewis*, 2020).

CASE 6.6 DNA ON FIREARM AND MULTIPLE CONTRIBUTORS

The defendant Kenneth Davon Lewis was charged with being a felon in possession of a firearm as an armed career criminal. Lewis challenged the admissibility of forensic DNA evidence relating to the firearm. He had a scuffle with police in the stairwell of an apartment building that occurred because Lewis was found at the property in violation of a court-restraining order. Lewis had just been released from prison and was allegedly at the property to collect some personal effects when someone spotted him in the stairwell and notified the police.

According to police when they confronted Lewis he was uncooperative and resisted arrest. During the encounter an officer saw something silver in Lewis's pocket and believed it was either a knife or a gun. The officer tried to handcuff Lewis but could not hold him or get him under control, and the owner of the building joined in the scuffle and saw a gun. Other officers also joined in and it was alleged the gun was ejected into the stairwell. Lewis claimed the gun was never in his possession, that he did not know its whereabouts and that it was planted in the stairwell to frame him. The court noted that evidence in the record established that police officers may have violated the lab's standard operating procedures by failing to wear gloves when handling the gun.

A DNA mixture profile found on the gun had a likelihood ratio 'greater than one billion times more likely if it originated from [Lewis] and three unknown unrelated individuals than if it originated from four unknown unrelated individuals'. All relevant police officers as well as the apartment manager were eliminated as contributors to the DNA mixture profile found on the gun. That calculation was offered to rebut Lewis's defence that, if his DNA was on the gun, it was transferred there by the police. Of the four assumed contributors to the mixture on the gun, the lab further determined that the contributor whose profile was consistent with that of Lewis was the major contributor, having contributed 56% of the total DNA recovered from the gun. The lowest presumed contributor, Contributor Number 4, contributed only 6% of the total DNA found in the mixture.

The report provided no further detail regarding the exclusion of the officers and apartment owner, including what, if any, likelihood ratio was generated to support that conclusion. The laboratory could not analyse more than a four-contributor mixture. The evidence at the hearing demonstrated that the laboratory had a high likelihood of underestimating the number of contributors to a DNA mixture. The Internal Validation Study of the laboratory was examined and demonstrated that, of the three analysts trained to perform the probabilistic genotyping, every lab analyst was more likely to underestimate the number of contributors than to get it right. The studies also established to the court that underestimating the number of contributors leads to false exclusions. Thus, while an underestimation of the number of contributors tends to have little effect on the likelihood ratio for inclusion, it tends to 'provoke' the exclusion of known contributors. The court ruling stated that the number of contributors in the case was deemed to be four, but given the lab's history, it was entirely possible, if not likely, that the true number of contributors was five or six.

The probabilistic genotyping system used was accepted, but the evidence regarding the exclusion of the officers as contributors to the DNA mixtures found on the gun was considered not reliable and was excluded. The court noted that studies have shown that Lewis's DNA could be on the gun even though he never handled it.

Case 6.6 can be compared with Case 6.5 with regard to the number of contributors. It is of interest that the validation studies of the laboratory were investigated at the hearing. It is always important to remember that the number of contributors is an estimate by the laboratory and a minimum. More sensitive techniques may elicit more contributors.

6.3.2 Levels of DNA Transfer

A discussion on different levels of transfer postulated to the trigger of a gun was in a section of a study on levels of DNA in a working forensic laboratory including case files (Taylor et al., 2016). The case had been analyzed by the particular authors.

A defendant yielded an inclusionary likelihood ratio when compared to a DNA profile obtained from a swab of the trigger of a gun that was found amongst clothing on a couch (ibid.). During court testimony a single primary transfer, hand to gun, was suggested by the prosecutor as the mode of transfer of the DNA analyzed.

However, it was stated in the paper that there are a myriad of secondary transfer pathways that exist, which would need to be considered in answering such an assertion:

- Hand to bag, bag to gun (*author note: unclear relevance of bag*).
- Body to clothes, clothes to gun.
- Body to couch, couch to gun.
- Defendant hand to offender hand, offender hand to gun, etc.

If considering tertiary transfer events then an even greater number of pathways exist, and this will be counterbalanced with a lower probability of detecting DNA from a tertiary transfer event. The combination of these two competing factors makes estimations about the likely number of transfer events a DNA result has originated from in real-life situations extremely difficult (ibid.).

The contaminations described in the body of the study (Taylor et al., 2016) were stated to provide some insight as out of 14 contaminations, only 9 could be explained by a primary event. The number of secondary and even tertiary transfer events resulting in detectable DNA was by no means insignificant. It is accepted that an initially deposited amount of DNA will be lost with each transfer step in a chain with each higher order of transfer. While a seemingly logical extension might suggest that most of the DNA we detect is from primary transfer events, this does not take into account the number of potential pathways and is in fact transposing evidence and propositions (ibid.).

The above reinforces the discussions in Chapter 3, whereby the analyst should not include the order of transfer in their propositions. Regarding activity levels, the propositions must be regarding activities and not levels of transfer.

6.4 Equipment and Implements

6.4.1 Persistence of DNA on Equipment and Implements

A paper from 2014 (van Oorschot et al., 2014) described the persistence of DNA deposited by the original user on objects after subsequent use by a second person. The conclusion was that the degree of persistence of DNA depends on the type of object, the substrate it is made of, the area of the object targeted for sampling and the duration and manner of contact by a subsequent user.

A Norwegian paper (Fonnelop et al., 2015b) studied the persistence and secondary transfer of DNA from previous users of equipment, such as a computer. The study showed that it was possible to transfer an initial user's DNA from a computer to the hands of a new user up to 8 days after receiving the equipment. Sampling between the keys of the keyboard identified the initial user's DNA to the end of the study, which was 42 days in length. This author notes that crevices and other cavities may retain DNA if unwashed and not touched and perhaps act like a reservoir of DNA, similar to bags and other personal possessions.

Background samples from the keyboards before the start of the experiment showed that in addition to the previous user, DNA from unknown persons was present (ibid.). This demonstrates the importance of sampling for 'background DNA' to assist in evaluating the persistence of trace DNA and the relevance of background or prevalent DNA in a specific case. See Chapter 4 for further discussion.

Examining car keys found away from a vehicle used by a temporary driver has been investigated (Boyko et al., 2019). The study found that DNA can be transferred from surfaces within a car to locations beyond the car by a temporary driver. Temporary drivers for a half-hour period exited a car that was usually driven by a 'regular user', placed car keys into a DNA-free dish, and then placed handprints on glass plates. The regular driver was a major contributor to profiles from some handprints, and this was explained due to the manner and duration of contact with the steering wheel for the period immediately preceding deposition of handprints.

The quantity of DNA retrieved from the car keys was up to 15.9 ng. One car key provided a mixed DNA profile of more than five persons. The regular driver was observed in all but one of the reportable profiles generated and was the major contributor on 12 occasions. The temporary driver was observed on 4 keys, on one occasion as a major contributor. These results were stated as not unexpected as the keys would have been handled by the temporary driver for only a relatively short period of time during the journey. DNA from a co-resident of a regular driver was observed in four key samples, twice as a major or co-major contributor. Unknown DNA sources were present in several samples.

6.5 Communal Items and Spaces

6.5.1 Direct and Indirect Transfer of DNA

If an item has been touched or otherwise had DNA deposited on it in communal spaces (direct transfer) it is likely to obtain a mixed DNA profile (van Oorshcot et al., 2019). The last 'user' is not necessarily the major contributor to any DNA profile obtained. This is because existing DNA in the environment may have been deposited onto a surface that the item contacted – indirect transfer. Or hands may contain non-self DNA that has been deposited through handling various items – again indirect transfer. See Chapter 4 for further discussion on non-self DNA on hands.

6.5.2 Public Spaces

One study (van den Berge et al., 2016) found that DNA from 51 public items such as railings at train stations, door handles and flush buttons of public toilets, handles of shopping carts and baskets, library books, banknotes and coins, was of useful quantity from the majority of samples with most having multiple contributors. Of the 51 banknotes and 6 coins analyzed where the last user was known, the last user was not necessarily the major contributor to the DNA profile obtained.

There needs to be a greater knowledge of probabilities of detection and relative contribution of individuals to DNA profiles from a wider array of shared objects and surfaces within shared spaces, particularly confined shared spaces such as offices and other work spaces. Forensic and police laboratories are discussed in Chapter 7 as these situations are crucial regarding examination due to the multiple exhibits from different cases coming together in one space. Such information assists in contamination mitigation practices.

6.5.3 Office and Work Spaces

Office spaces may be frequented by the normal work personnel of that space, as well as visitors. The predominance of open-plan offices with multiple people occupying the same area may result in direct DNA deposits on office surfaces from many people at many times, as well as indirect deposits. Newsrooms, phone call centres, financial markets (e.g., 'Wall Street') have multitudes of people crammed into shared spaces. Even large offices occupied by senior personnel may also be occupied by assistants, visitors, cleaners and persons performing senior duties while the usual occupier is on vacation. Thus DNA from multiple persons is expected on surfaces and items within these spaces.

An Australian/Dutch study investigated to what extent DNA left by a temporary user of an office space, occupied by a regular user for an extended period, is detectable when the duration of the temporary occupancy and general activities is known and, further, how readily the DNA of the regular user is still detectable after a known period of occupancy by another person and to what extent DNA of others is present (Goray et al., 2020). Eight single-use office spaces within two forensic laboratories (on different continents) had a temporary user for 2.5 hours to 7 hours. There were 18 core items/surfaces that were examined. The owner/

regular user of the office space was found to be the major contributor to DNA profiles from most items within the office space, but the detection of the temporary user of the office space varied among offices and items (ibid.). A difference in some methodologies between the two laboratories (and countries on different continents) appeared to impact DNA yields and DNA profile types. These variables are what is used in 'activity level' assessments, showing the problem in applying data from literature to case scenarios in a particular jurisdiction (this author opinion). Also note the varying degree of pre-examination and sampling by police in jurisdictions before submission for DNA examination, which also introduces unknown variables.

A large variation was observed in DNA yields between item surfaces no matter in what laboratory the experiment was performed (ibid.). The highest yields of DNA were obtained from the chair seats and computer keypads (laboratory A) and chair back rests and computer keypads (laboratory B). This author has already noted these types of items may be reservoirs of DNA.

An interesting finding was that the number of contributors ranged from 1 to 5 in laboratory A and from 1 to 3 in laboratory B. A majority of the samples profiled in laboratory A (90% of interpretable profiles) and laboratory B (76% of interpretable profiles) had more than one contributor to the DNA profile generated (ibid.). The same DNA profile interpretation system and casework methods were used by both laboratories – that used by laboratory A. Individuals were reported as having contributed to a sample if a likelihood ratio is greater than one but by itself 'does not necessarily mean that the individual in question contributed DNA to the sample in question' (ibid.). It is interesting also to note that this was explained as due to semi-controlled experiments where DNA from non-core participants may be present in the sample. It is possible that this refers to other laboratory members or their relatives, due to a likelihood ratio of 10,000 being marked with an asterix; although not explained this may be the threshold for staff DNA adventitious matches. This author queries the specificity of the interpretation at low likelihood ratios – if adventitious matches of likelihood ratios of less than 10,000 may occur to staff, then they may occur to anyone.

Non-office owner DNA, or temporary occupier DNA, was observed in 82% of profiles for laboratory A and 65% of profiles for laboratory B. DNA from the known domestic partner of the owner of one office was identified as a contributor on the chair back rest and chair lever, and on the chair lever in another office. Various work colleagues were identified as contributors on eight occasions for laboratory A, and on 14 occasions for laboratory B. Unknown source DNA was found in 75% of samples from laboratory A and 39% of samples for laboratory B.

References

Bowman, Z., Mosse, K., Sungalla, A., et al., 2018, Detection of offender DNA following skin-to-skin contact with a victim, *Forensic Science International: Genetics*, 37, 252–259.

Boyko, T., Kokshoorn, B., and van Oorschot, R., 2019, Prevalence of DNA in vehicles: Linking an item away from a vehicle to occupancy of the vehicle, *Forensic Science International: Genetics Supplement Series*, 7, 829–830.

Breathnach, M., Williams, L., McKenna, L., et al., 2016, Probability of detection of DNA deposited by habitual wearer and/or the second individual who touched the garment, *Forensic Science International: Genetics*, 20, 53–60.

Buckingham, A., Harvey, M., and van Oorschot, R., 2016, The origin of unknown source DNA from touched objects, *Forensic Science International Genetics*, 25, 26–33.

Buckingham, A., Harvey, M., and van Oorschot, R., 2017, Transfer of picked-up DNA to cotton plates, *Forensic Science International: Genetics Supplement Series*, 6, e6–e8.

Burrill, J., Daniel, B., and Frascione, N., 2018, A review of trace "touch DNA' deposits: Variability factors and an exploration of cellular composition, *Forensic Science International: Genetics*, 39, 8–18.

Butcher, E., van Oorschot, R., Morgan, R., et al., 2019, Opportunistic crimes: Evaluation of DNA from regularly-used knives after a brief use by a different person, *Forensic Science International: Genetics*, 42, 135–140.

Cale, C., Earll, M., Latham, K., et al., 2016, Could secondary DNA transfer place someone falsely at the scene of a crime? *Journal of Forensic Sciences*, 61, 1, 196–203.

Casey, D. and Price, J., 2010, The sensitivity and specificity of the RSID(TM)-saliva kit for the detection of human salivary amylase in the Forensic Science Laboratory, Dublin, Ireland, *Forensic Science International*, 194, 67–71.

De Wolff, T., Aarts, L., van den Berge, M., et al., 2019, Prevalence of DNA in vehicles: Linking clothing of a suspect to car occupancy, *Australian Journal of Forensic Sciences*, 51, S1, S103–S106.

Fonnelop, A., Egeland, T., and Gill, P., 2015a, Secondary and subsequent DNA transfer during criminal investigations, *Forensic Science International: Genetics*, 17, 155–162.

Fonnelop, A., Johannessen, H., and Gill, P., 2015b, Persistence and secondary transfer of DNA from previous uses of equipment, *Forensic Science International: Genetics Supplement Series*, 5, e191–e192.

Fonnelop, A., Johannessen, H., Egeland, T., et al., 2016, Contamination during criminal examination: Detecting police contamination and secondary DNA transfer from evidence bags, *Forensic Science International: Genetics*, 23, 121–129.

Fonnelop, A., Ramse, M., Egeland, T., et al., 2017, The implications of shedder status and background DNA on direct and secondary transfer in an attack scenario, *Forensic Science International: Genetics*, 29, 48–60.

Gill, P., 2016, Analysis and implications of the miscarriages of justice of Amanda Knox and Raffaele Sollecito, *Forensic Science International: Genetics*, 23, 9–18.

Gill, P., Hicks, T., Butler, J. et al., 2018, DNA commission of the international society for forensic genetics: Assessing the value of forensic biological evidence – Guidelines highlighting the importance of propositions, Part I: evaluation of DNA profiling comparisons given (sub)-source propositions, *Forensic Science International: Genetics*, 36, 189–202.

Gill, P., Hicks, T., Butler, J., et al., 2020, DNA commission of the international society for forensic genetics: Assessing the value of forensic biological evidence - Guidelines highlighting the importance of propositions. Part II: Evaluation of biological traces considering activity level propositions, *Forensic Science International: Genetics*, 44, 1–13.

Gosch, A. and Courts, C., 2019, On DNA transfer: The lack and difficulty of systematic research and how to do it better, *Forensic Science International: Genetics*, 40, 24–36.

Gosch, A., Euteneuwer, J., Preuss-Wössner, J., et al., 2020, DNA transfer to firearms in alternative realistic handling scenarios, *Forensic Science International Genetics*, 48, 102355.

Goray, M., Van Oorschot, R., and Mitchell, J., 2012, DNA transfer within forensic exhibit packaging: Potential for DNA loss and relocation, *Forensic Science International: Genetics*, 6, 158–166.

Goray, M., Kokshoorn, B., Steensma, K., et al., 2020, DNA detection of a temporary and original user of an office space, *Forensic Science International: Genetics*, 44, 102203.

Hicks, T., Buckleton, J., Castella, V., et al., 2022, A logical framework for forensic DNA interpretation, *Genes*, 13, 957.

James, M., 2021, Degrees of contrast: Detection of latent bloodstains on fabric using an alternate light source (ALS) and the effects of washing, *Journal of Forensic Sciences*, 66, 1024–1032.

Jones, S., Scott, K., Lewis, J., et al., 2016, DNA transfer through non-intimate social contact, *Science and Justice*, 56, 90–95.

Kafarowski, E., Lyon, A., and Sloan, M., 1996, The retention and transfer of spermatozoa in clothing by machine washing, *Canadian Society Forensic Science Journal*, 29, 1, 7–11.

Kulstein, G. and Wiegand, P., 2018, Comprehensive examination of conventional and innovative body fluid identification approaches and DNA profiling of laundered blood and saliva stained pieces of cloths, *International Journal of Legal Medicine*, 132, 1, 67–81.

McKoll, D., Harvey, M., and van Oorschot, R., 2017, DNA transfer by different parts of a hand, *Forensic Science International: Genetics Supplement Series*, 6, e29–e31.

Meakin, G., Butcher, E., van Oorschot, R., et al., 2015, The deposition and persistence of indirectly-transferred DNA on regularly-used knives, *Forensic Science International: Genetics Supplement Series*, 5, e498–e500.

Meakin, G., Butcher, E., van Oorschot, R., et al., 2017, Trace DNA evidence dynamics: An investigation into the deposition and persistence of directly and indirectly transferred DNA on regularly used knives, *Forensic Science International: Genetics*, 29, 38–47.

Murphy, C., Kenna, J., Flanagan, L., et al., 2020, A study of the background Levels of male DNA on underpants worn by females, *Journal of Forensic Sciences*, 65, 399–405.

Noel, S., Lagace, K., Rogic, A., et al., 2016, DNA transfer during laundering may yield complete genetic profiles, *Forensic Science International: Genetics*, 23, 240–247.

Ogawa, H., Hiroshige, Y., Yoshimoto, T., et al., 2018, STR-genotyping from a dry-cleaned skirt in a sexual assault case, *Journal of Forensic Sciences*, 63, 4, 1291–1297.

Otten, L., Banken, S., Schurenkamp, M., et al., 2019, Secondary DNA transfer by working gloves, *Forensic Science International: Genetics*, 43, 102126.

Ramos, P., Handt, O., and Taylor, D., 2020, Investigating the position and level of DNA transfer to undergarments during digital sexual assault, *Forensic Science International: Genetics*, 47, 102316.

Reither, J., Gray, E., Durdle, A., et al., 2019, Background DNA on flooring: The effect of cleaning, *Forensic Science International: Genetics Supplement Series*, 7, 787–790.

Reither, J., Gray, E., Durdle, A., et al., 2021, Investigation into the prevalence of background DNA on flooring within houses and its transfer to a contacting surface, *Forensic Science International*, 318, 110563.

Reither, J., van Oorschot, R., and Szkuta, B., 2022, DNA transfer between worn clothing and flooring surfaces with known histories of use, *Forensic Science International: Genetics*, 61, 102765.

Ruan, T., Barash, M., Gunn, P., et al., 2018, Investigation of DNA transfer onto clothing during regular daily activities, *International Journal of Legal Medicine*, 132, 1035–1042.

Samie, L., Taroni, F., and Champod, C., 2020, Estimating the quantity of transferred DNA in primary and secondary transfers, *Science and Justice*, 60, 128–135.

Sari, D., Hitchcock, C., Collins, S., et al., 2019, Amylase testing on intimate samples from pre-pubescent, post-pubescent and post-menopausal females: Implications for forensic casework in sexual assault allegations, *Australian Journal of Forensic Sciences*, 52, 6, 618–625.

Szkuta, B., Harvey, M., Ballantyne, K., et al., 2015, DNA transfer by examination tools- a risk for forensic casework? *Forensic Science International: Genetics*, 16, 246–254.

Szkuta, B., Ballantyne, K., and van Oorschot, R., 2017, Transfer and persistence of DNA on the hands and the influence of activities performed, *Forensic Science International Genetics*, 28, 10–20.

Szkuta, B., Ansell, R., Boiso, L., et al., 2019, Assessment of the transfer, persistence, prevalence and recovery of DNA traces from clothing: An inter-laboratory study on worn upper garments, *Forensic Science International: Genetics*, 42, 56–68.

Szkuta, B., Ansell, R., Boiso, L., et al., 2020, DNA transfer to worn upper garments during different activities and contacts: An inter-laboratory study, *Forensic Science International: Genetics*, 46, 102268.

Taylor, D., Abarno, D., Rowe, E., et al., 2016, Observations of DNA transfer within an operational forensic biology laboratory, *Forensic Science International: Genetics*, 23, 33–49.

Taylor, D., Biedermann, A., Hicks, T., et al., 2018, A template for constructing Bayesian networks in forensic biology cases when considering activity level propositions, *Forensic Science International: Genetics*, 33, 136–146.

USA v Kenneth Davon Lewis [2020] District court of Minnesota, Case No. 18-cr-194 (ADM/DTS).

van den Berge, M., Ozcanhan, G., Zijljstra, S., et al., 2016, Prevalence of human cell material: DNA and RNA profiling of public and private objects and after activity scenarios, *Forensic Science International: Genetics*, 21, 81–89.

van Oorschot, R., Glavich, G., and Mitchell, J., 2014, Prevalence of DNA deposited by the original user on objects after subsequent use by a second person, *Forensic Science International: Genetics*, 8, 1, 219–225.

van Oorschot, R., Szkuta, B., Meakin, G., et al., 2019, DNA transfer in forensic science: A review, *Forensic Science International: Genetics*, 28, 140–166.

Willis, S., McKenna, L., McDermott, S., et al., 2015, *ENFSI guidelines for evaluative reporting in forensic science*, European Network of Forensic Science Institutes, available at http://enfsi.eu/documents/forensicguidelines/

Chapter 7

Inadvertent DNA Transfer

BOX 7

- Quality procedures and documentation.
- Personal protection equipment.
- Crime scene.
- The mortuary.
- Medical examination.
- Packaging, handling and storage.
- Police and other laboratories.
- The forensic biology laboratory.
- Staff DNA elimination databases.

7.0 Introduction

Forensic DNA transfer may be a contamination risk and compromise the quality and integrity of the final result, as described in the previous chapters. This author describes this as 'inadvertent DNA transfer'.

Addition of DNA to a sample collected after the suspected event can complicate the interpretation of a DNA profile result/s, and/or misdirect criminal investigations. It may even lead to flawed

DOI: 10.4324/9781003158844-7

convictions and miscarriages of justice. Cases of inadvertent DNA transfer include those where the accused was imprisoned for rape for 14 months when there was no offence at all (Case 1.3 Farah Jama), a wide disparity of offences including murder and burglary by different people linked by one contaminating person that misdirected investigations (Case 3.1 Phantom of Heilbronn), and even a forensic laboratory staff member accused of murder and rape leading to his suicide (Case 2.5 Kevin Brown).

DNA contamination and DNA transfer are the same physical phenomena of the movement of DNA from one surface or location to another. It is the timing of the movement that defines whether DNA transfer could be associated with a crime-related activity prior to securing the crime scene – or a non-crime-related contamination event during or post securing a crime scene.

This author refers to the addition of DNA after the alleged crime event as 'inadvertent transfer' of DNA in the context of contamination and quality failure. This is usually investigator-mediated, but may also occur due to contamination of reagents and containers that the investigator relies upon to be 'DNA free'. The examiner may be unaware that they are a mediator of trace DNA to a crucial item of evidence.

Crime-related DNA transfer may refer not only to the offender transferring their own DNA to an object but also to an offender transferring other (non-self) DNA to the crime scene. This can be considered as 'inadvertent transfer' as the offender may not be aware that they have transferred DNA that is not their own. Note that this may associate a person with a crime when they have not been there and is a main area of concern regarding domestic and social acquaintances. This book has discussed many instances in the literature of controlled studies in a laboratory where the DNA of domestic and social contacts of participants (and even 'unknown people') have been detected on items examined in the studies.

Non-crime-related contamination can arise through many avenues (or vectors) such as a police officer at a scene, a scientist examining the evidence, a dirty examination tool or a non-DNA-free reagent used during sample analysis (van Oorshcot et al., 2019).

This book author worked in forensic laboratories and was aware of current practices in other laboratories, before the ready transfer of trace DNA was appreciated, prior to and during the first decade of this century. Laboratory coats were changed weekly or at the discretion of the wearer, masks or hair nets or overshoes were not worn, gloves were considered as a protection for the examiner but nothing else, environmental or background DNA monitoring was not in place and examiners worked on multiple cases (even if sequentially) at the one bench – with these benches not identified in case notes.

Examination at crime scenes and even at the mortuary had proceeded on the basis of protection of the examiner from the environment. It is only within the past decade that it has been realized that personal protection equipment (PPE) such as gloves may be efficient transfer vectors for DNA. Although procedures may have changed in laboratories to account for the prevention of inadvertent transfer, these may not necessarily have been adopted by police or other personnel.

Recent literature has noted that with advances in DNA technology and detection capability, there is also an increased probability of detecting trace DNA from contamination, and in DNA from higher-order transfers such as tertiary transfer. The risk of detecting DNA that has been transferred to exhibits during the forensic process, and is not related to an alleged offence, has also subsequently increased. Research within operational forensic DNA laboratories may provide insight into the possible mechanisms that lead to exhibit contamination (Mercer et al., 2022). Additionally, contaminating DNA can create a mixed DNA profile or mask the offender's profile from a sample, which can decrease the evidential value of a 'match' with a person of interest or result in the loss of information that could have been used to identify an individual (ibid.).

Note that police and forensic DNA laboratories are where multiple exhibits from multiple cases may co-exist and potentially comingle, providing an excellent opportunity for inadvertent transfer of DNA.

7.1 Importance of Quality Procedures, Documentation and Contamination Mitigation

Contamination with DNA depends on the availability and opportunity of DNA transfer to the item in question. Some ways to detect this include an assessment of the scientific interpretation (part of which includes an independent review), exploration of continuity and staff DNA database checks, all of which also rely on proper documentation. A review of error rates in the laboratory may also indicate specific or global problems in the laboratory (Kloostermann et al., 2014).

Contamination may be introduced by investigators at any point and thus it follows that the entire investigative process should be open to scrutiny and best practice procedures should be followed from the beginning to the end of reporting the evidence (Fonnelop et al., 2016). DNA contamination incidents are one of the most frequent sources of error in forensic genetics and can have serious consequences (Bassett and Castella, 2019).

There is no specific 'test' for contamination. Audit trails may follow the path of the exhibit from collection to final analysis and indicate areas in which contamination may be a possibility. Other evidence may indicate that it was physically impossible for the accused to have committed the crime –such as being in prison at the time, in a hospital (Case 4.1 Lukis Anderson) or overseas.

Continuity and audit trails can assist in querying the potential for error at each step. An audit trail of a sample should have a unique identifier that enables the sample to be followed at each stage of the analysis so that questions such as which person handled the sample on a date and at a location can be readily answered.

Contamination may be sporadic and not 'global' or systemic (United Kingdom Forensic Science Regulator, 2016, Guidance 206). It may not be known how, or if, contamination in a matter may have occurred.

It is important to obtain as much information as possible regarding the history of the item that is examined for any subsequent DNA profiling. It is not possible to determine from a DNA profile when the DNA was deposited or its 'age', and any DNA profile must be viewed in context. The time that the DNA was deposited may be the ultimate issue. How the DNA got on the exhibit, and whether that DNA is relevant to the offence, may also be in question. The forensic examiner should be assured that the sample that they are examining has been obtained in a manner following quality assurance procedures.

The following two cases illustrate a principle of quality assurance that is recognized by the judicial system, and the requirements of 'burden of proof'. These cases highlight the principles of trace evidence examination and the need for continuous quality from the moment of seizure to the analysis of the DNA trace. The quality of the final result (for example, the DNA profile) depends on the quality at each step of the analysis.

The first case concerned a bombing in Northern Ireland where poor forensic practices raised serious questions (*R v. Hoey*, 2007). The trial received high publicity in the United Kingdom, and the trial verdict of an acquittal damning of the crime scene and laboratory practices created much concern. The judicial comments are still relevant today – the comments described below are quite lengthy but important. The court ruling is alarming reading and encompasses failures from the collection, handling, storage and packaging that are graphically illuminating. Some of the failures have also arisen in later years in other countries in other trace forensic DNA cases and thus are notable as they are seemingly persistent, particularly disturbing in tragic cases impacting many people as the following demonstrates.

CASE 7.1 OMAGH BOMBING (*R V. HOEY*, 2007)

The devastating Omagh bombing occurred in 1998 during which 29 people were killed and 220 injured in a car bomb attack in Northern Ireland, suspected to be committed by the 'Real IRA'. Sean Hoey was charged in 2005 after it was alleged that his DNA was found on bomb timers collected during the crime scene examination. Trace DNA was collected that was at low levels and the technique of 'low copy number' (higher amplification cycles than the current routine of 28 cycles) was used. Justice Weir stated: 'At the time that they were being recovered the possibility of their being examined for DNA was not considered by those effecting their recovery, storage and transmission. It is said that the result of DNA examination is to show that the accused was in contact with parts of the devices'.

Multiple instances of poor contamination mitigation were exposed during the trial. Two officers claimed to be wearing personal protection suits but were shown in photographs to not be wearing any, and their proclaimed awareness of the ready transfer of DNA at the time was at odds with all the other investigators in the case, causing their evidence to be dismissed.

One scenes of crime officer 'handed items over to an exhibits officer for the investigation but he could not say who that was as they did not at that time make records as to when and to whom items had been handed over. He said that at the time the (name) special property store was "a complete mess", that often an item he was looking for within it had not been recorded in the book and that bags within it could "spring a leak". He concluded on this topic by saying "fortunately I wasn't in the special property store very often, it was a mess"' (ibid.).

'The last witness described the procedure he followed on being sent there to collect items. He said (the forensic laboratory) would have them ready on his arrival. The items would be brought out in something like a supermarket trolley, maybe up to four or five trolley loads. He would then go back to reception and sign. He trusted them that they had given him the forms to which the items related to sign. There was therefore no system for verifying what had been collected or what was then placed in the police store' (ibid.).

Within the forensic examining laboratory there were also problems. When the police realized it was possible to further investigate they wanted to 'discover the whereabouts of and gather in the exhibits from Omagh and what were believed to be the various linked scenes. Labels had become detached from items because they were held on with sellotape which had aged and dried up. (The forensic scientist) agreed that this had led to a problem in another prosecution where DNA findings had been wrongly attributed to an accused and that this and an "accumulation of minor errors, minor ineffectiveness of the quality system" had led to the temporary suspension of the laboratory's accreditation by the United Kingdom Accreditation Service ... a memo written by her to a colleague during the search: "Please search this week, the SFU garage, photography, Bio boxes for any unidentified bits that might be related to the Omagh or related incidents i.e. anything lying around without a label". She explained the purpose of this by saying: "all we were trying to do was spring clean those areas and make sure there was nothing identifiable from any of the incidents we were looking at." When I asked what would happen if items were recovered either from within or without the laboratory whose prior whereabouts or the nature of any handling in the interim were unknown her response was that if the item were recovered it would be assessed as to its condition and the handling that it had had before a decision was made as to whether further work could be done on it. How this assessment was to be usefully made without reliable information as to where the item had been, with whom and in what conditions was not explained by the witness' (ibid.).

The laboratory scientist 'was asked about the laboratory practice in relation to allowing persons not involved in the examination of items to look at them. Her initial response was that if they did so they would in general be looked at through the packaging. She was then asked about an e-mail to her ... in the following terms: "Subject: Omagh "I was looking at the minutes of the last Forensic Management Team meeting. I don't know where profiles/partials are coming up in the Omagh case but I know that I was shown the bit of TPU box out in Explosives. *I don't remember touching anything but who knows. Lots of other people were shown these things although I guess most would not have had actual contact. Were there any profiles found on pieces recovered from the debris? – If so I presume the profiles of the recovery team have all been eliminated'.*

Justice Weir stated: 'This is a disturbing document, contemplating as it does at least the real possibility that an item that was subject to LCN DNA examination had previously been handled by (laboratory director) and even more seriously, by other persons whose identity was not known .... My subsequent description of some of the evidence concerning the actual regime that prevailed both within the police and (laboratory) during the relevant period demonstrates how far short of surmounting that high hurdle the prosecution has fallen in this case. It is not my function to criticize the seemingly thoughtless and slapdash approach of police and SOCO officers to the collection, storage and transmission of what must obviously have been potential exhibits in a possible future criminal trial but it is difficult to avoid some expression of surprise that in an era in which the potential for fibre, if not DNA, contamination was well known to the police such items were so widely and routinely handled with cavalier disregard for their integrity'.

Justice Weir stated: 'The evidence establishes that the arrangements within the police in 1998 and 1999 for the recording and storage of items were thoroughly disorganized. There were numerous examples of this during the trial with labels missing from items or incorrectly attached to the wrong item. Examples were given of labels lying loose and bags without labels. There was no universal system of logging items received, no proper recording in police stations so no inventory of what was in a store at any particular time …

What I do find extraordinary is that, knowing that these items had not been collected or preserved using methods designed to ensure the high degree of integrity needed not merely for DNA examination but for the more exacting requirements of LCN DNA, examinations were performed at Birmingham with a view to using them for evidential rather than solely intelligence gathering purposes. The findings of those examinations were put forward and stoutly defended by (name) of the Birmingham FSS laboratory as evidence that the Court might safely rely upon as tending to establish the guilt of the accused'.

The court 'must be satisfied by the prosecution witnesses and supporting documents that all dealings with each relevant exhibit have been satisfactorily accounted for from the moment of its seizure until the moment when any evidential sample relied upon by the prosecution is taken from it and that by a method and in conditions that are shown to have been reliable. This means that each person who has dealt with the item in the intervening period must be ascertainable and be able to demonstrate by reference to some proper system of bagging, labelling, and recording that the item has been preserved at every stage *free from the suspicion of interference or contamination*. For this purpose they must be able to demonstrate how and when and under what conditions and with what object and by what means and in whose presence he or she examined the item' (Italics are by this current author).

A recently formed dissident republican group, calling itself the Real IRA, had claimed responsibility for the bomb in 1998 when in a statement the paramilitary group said its targets were 'commercial' and offered an apology to the 'civilian' victims (BBC News, 2021). Investigations and numerous court hearings continued after the above verdict, with no successful criminal trial to date. The victims' families begin a landmark civil case in 2008, suing five men they alleged were involved from the now-defunct Real IRA. The case broke new legal ground and is believed to be the first time anywhere in the world that alleged members of a terrorist organization have been sued (ibid.). Four persons were convicted – Colm Murphy, Seamus Daly, Michael McKevitt and Liam Campbell (McDonald, 2013).

There is a message for forensic scientists here that defend their results within the prism of their own laboratory and do not explore the history of the exhibit that they have received, which is not a whole of case perspective (discussed in Chapter 1). The Australian case of Farah Jama (Case 1.3) also concerned forensic scientists proclaiming the reliability of their results and the absence of contamination when queried by detectives. Justice Vincent stated that a crucial moment in the investigation of the case, which effectively set the stage for what followed, was when the investigating detective asked about instances of contamination in the forensic laboratory and the response of the case reporting scientist was 'in my opinion I do not think that contamination between the cases could have occurred as items from the two cases were examined at different times, at different areas and by different people'. The statement was also incomplete as records showed that another staff member examined the swabs and slides from the first complainant 4 days earlier than the swabs and slides from the second complainant at the same bench (Vincent, 2010). The Vincent report stated that at a minimum the limitations upon the opinion expressed should have been given at the outset which may have prompted a broader enquiry (ibid.).

Absolving oneself from responsibility when examining an item calls into question why an examination was performed in the first place without checking the previous processes, and when the DNA reporting scientist is the one with the expertise regarding DNA contamination mitigation. The reporting scientist may also be confronted with video footage in court.

Another issue raised by Justice Weir (*R v. Hoey*, 2007) in Case 7.1 was the debate concerning low copy number and validation, with his finding that it had not been validated. The concerns expressed are still relevant today, especially with ever more sensitive techniques applied to ever more samples, and new ways of interpreting DNA evidence such as under 'activity level' propositions. Justice Weir stated: 'Validation' is defined in guidelines as 'the process whereby the scientific community acquires the necessary information to:

- Assess the ability of a procedure to obtain reliable results.
- Determine the conditions under which such results can be obtained.
- Define the limitations of the procedure".

The second highly publicized case has been discussed previously (Case 3.2).

CASE 7.2 MURDER OF MEREDITH KERCHER (HELLMANN, 2011)

An Appeal hearing in Italy in 2011 occurred after the first conviction of Amanda Knox and Raffaele Sollecito for the murder of Meredith Kercher in Italy, which ultimately acquitted them (prior to a further trial and conviction). Issues revolved around trace DNA and particularly low levels of DNA.

An argument had been put by the prosecution that it was not enough for the defence to say that the DNA result was from contamination – the burden was on those claiming contamination to prove its origin. However, the Appeal Court held that the 'burden' showed the result was obtained using a procedure which guaranteed the integrity of the item, from the moment of collection to the moment of analysis. Once there is no proof that these precautions have been taken then it is not necessary to also prove the specific source of the contamination (Hellmann, 2011).

7.2 Personal Protection Equipment

Adequate personal protection equipment should be mandatory for personnel handling exhibits subject to DNA examination. This refers to the collection at the crime scene to final examination at the forensic biology laboratory. Such equipment includes gloves, a mask, a protective over suit or overalls, coverings for shoes and a hairnet.

The wearing of personal protection equipment is necessary for quality assurance when handling exhibits for trace evidence. Decades ago protection equipment was designed to protect the wearer from hazardous items and the environment at the forensic biology crime scene. Masks and other breathing equipment were only used for known hazardous aerosols, such as when using luminol spray for possible blood detection. Gloves were used at crime scenes with biological evidence to protect the wearer from biological hazards such as contaminated blood – AIDS and Hepatitis B were a particular issue in the 1980s and early 1990s and gloves were mandatory for user protection (author information from working at crime scenes during these decades).

During the 2000s some countries such as the United Kingdom implemented full crime scene personal protection equipment (PPE) designed not only to protect the wearer but also to protect the environment and the exhibits collected from cross-contamination (author information from working in the UK). These procedures evolved from contamination minimization protocols for fibre transfer and were extrapolated to trace DNA analysis. A familiar sight on news TV screens in the United Kingdom was crime scene personnel in white suits examining the crime scene, in stark contrast to USA crime TV shows that depicted glamorous detectives in designer outfits wandering through crime scenes unprotected.

The use of gloves, out of all personal protection equipment, has been a particular area of research due to gloves being worn on the hands, acting as a vector or means of transmission of DNA. Other protection equipment would benefit from further study, particularly overshoes when 'stepping plates' are not used at crime scenes.

Disposable gloves can act as an efficient transfer vector between items as illustrated in a Norwegian study in regards to tertiary transfer of DNA (Fonnelop et al., 2015), where gloves acted as a vector from one item to another. An Australian study demonstrated that DNA material can be transferred from exhibit to exhibit by scissors, forceps and gloves (Szkuta et al., 2015). Gloves were noted as more efficient as a transfer vector than scissors or forceps. These instruments pose a significant contamination risk if not DNA free before contact is made with the targeted sample during exhibit examination. The reuse of instruments including gloves and further contact with other areas of an exhibit could potentially relocate DNA, which could negatively affect the interpretation of relevant activities. The potential of intra-exhibit transfer is greater with trace DNA samples as they are not visible.

The study (Szkuta et al., 2015) recommended the following:

- Using disposable forceps and scissors.
- Cleaning of gloves with the appropriate agent before use.

- Changing of gloves each time after touching the surface of an exhibit.
- Wearing multiple gloves to reduce skin exposure when changing gloves and
- Reduce handling of items if later sampling for DNA.

Although disposable clothing and gloves may prevent the investigator's body from contaminating the crime scene, there is still the potential for the investigator to act as an unwitting vector of DNA within the crime scene itself.

Gloves need to be changed frequently, not only between each item but including between different areas of the same item (such as between the blade and handle of a knife). Cleaning of gloves before use and wearing multiple layers of gloves has been recommended (Margiotta et al., 2015; Szkuta et al., 2015) to avoid skin exposure during the changing of gloves.

Gloves worn during exhibit examination are a key risk for contamination of exhibits (Goray et al., 2019). The risk will depend on how gloves are put on, what they touch and when they are replaced. Addition, loss and/or redistribution of DNA within, and to and from, an exhibit through handling during an examination can have significant implications for the investigation, placing doubt on the meaning of any results obtained.

A Norwegian study (Fonnelop et al., 2016) also noted that if the correct handling procedures are not followed the mere wearing of the recommended protection equipment is insufficient, giving a false sense of security. Recommendations include:

- Victim and suspect exhibits kept separate.
- All staff in contact with exhibits wear mask, hair net, gloves, clean lab coats or disposable investigation suits.
- Gloves changed before and after touching exhibit and consider double gloving.
- Evidence bags should be handled as little as possible and
- Results from old 'unsolved' cases where the continuity is unclear or carried out using procedures that pre-date the era of high-sensitivity profiling, may be irretrievably compromised. The interpretation of such evidence must be very cautious.

Note the last point in regards to 'cold cases', exemplified by Case 7.1.

7.3 Crime Scene

7.3.1 Collection of Exhibits

The first step of the examination path of an exhibit is the crime scene, whether that be the physical location in which the event occurred or on the body of a person, such as an autopsy or a medical examination.

This section discusses the physical location of the crime scene/s in which the alleged crime event took place. DNA may be all pervasive in the environment due to those who inhabit that environment, constantly shedding and transferring self-DNA and also non-self-DNA. Background DNA on surfaces may be present for years provided that the environment is dry and undisturbed (see Chapter 4).

While crime-associated DNA transfer occurs only before the crime scene itself is established by authorities, contamination can only occur afterwards. Newer DNA testing kits have increased awareness of the ready transfer of detectable DNA in forensic DNA laboratories but this has often not translated to police personnel (Fonnelop et al., 2016).

Investigator-mediated contamination (Gill, 2014) may occur inside the crime scene but mediated by investigators who unwittingly (inadvertently) contaminate the crime scene with their own DNA, or act as a mediator with their gloves or other personal equipment ('investigator mediated transfer').

Appropriate training and increasing awareness of policies surrounding, for example, the use and replacement of personal protection equipment can assist in preventing contamination. Methodological changes may improve detection of extraneous DNA, while inclusion of investigators and police officers in a DNA elimination database assist in detection following its occurrence.

The UK Forensic Science Regulator in 2016 (Forensic Science Regulator, 2016) issued guidelines regarding the control and avoidance of contamination at crime scenes, Guidance 206. Contamination incidents cannot be eliminated completely given the prevalence of human DNA within the living and working environment. This issue is exacerbated by the increasing sensitivity of DNA analytical techniques.

Section 6.1.20 of this Guidance notes that there are contamination risks between different parts of the same scene such that a change of barrier clothing and different personnel should be used. Barrier clothing should be retained such as gloves, overalls and foot coverings.

Section 8.1.2 of this Guidance suggests steps to minimize the chance those handling the items may cause inadvertent DNA contamination and references the paper (Fonnelop et al., 2016) mentioned above.

If an object is located on a surface at a crime scene, 'control' samples should be taken off the surface to assess the presence of the 'background' level of DNA (Fonnelop et al., 2017). Control samples are further discussed in this chapter.

7.3.2 Photographs and Video Recordings

Photographs and video recordings may be useful in depicting the exact use of personal protection equipment at the scene and in handling exhibits. Both Case 7.1 (photographs) and Case 7.2 (video recordings) showed flawed collection practices when collecting exhibits for trace DNA.

A bizarre case from the author's files had video recordings of the seizure of drugs and a firearm – however the videos were selective and concerned the episode where the accused was taken to the scene and showed police 'discoveries'. The videos showed multiple flawed collection practices. An appeal had been launched in the Supreme Court after the conviction of the defendant and the author was asked to review the case.

CASE 7.3 TRACE DNA ON FIREARM

A suspect was arrested driving his car by two police officers as part of a joint drugs task force. One police officer was designated as the crime scene officer and exhibit manager and was from the national crime police force. The other police officer was from the State police and was designated as the scenes of crime photographer and video recorder. The two police officers were twins.

The suspect was taken to the scene by the twins in their police vehicle. The scene was a large area of outback land and there were locations where 'drugs and firearms' had been buried and subsequently located by the police operation. A firearm had been previously located and it was covered in a towel. Videos showed the suspect taken to the firearm location, drinking from a water bottle and having a cigarette along the way. He was shown the unwrapping of the firearm from the towel by the crime scene officer. DNA found on the firearm was proposed to link the suspect to the firearm.

There were numerous instances captured on the video recording showing improper contamination mitigation procedures, including infrequent changes of gloves and glove boxes and crime scene kits adjacent to exhibits, crime scene personnel wearing jeans and tops and not protective overalls or masks, and a general cavalier disregard for the opportunity for trace DNA transfer with the suspect at the scene. The suspect was wearing normal street wear.

The appeal reduced the term of imprisonment.

Video recordings from the crime scene collection in a case from Australia eventuated in a successful appeal, overturn of conviction and the immediate release of the defendant from prison. Evidence against the accused in the original trial was based on five circumstantial matters and the prosecution accepted that only in its entirety could the jury establish guilt beyond reasonable doubt (*Seifeddine v. R*, 2021).

CASE 7.4 DNA ON TRIGGER OF A PISTOL

The appellant was a part-time chef and was charged on a joint indictment with the café owner with counts of firearms and prohibited weapons offences based on their alleged possession. After a trial jury in 2019 both men were convicted and Mr. Seifeddine sentenced to 5 1/2 years prison. The critical evidence against him was the presence of DNA matching his DNA profile on the trigger area of a Smith and Wesson revolver. His contention on appeal was that the DNA evidence was compromised by police, and there was a reasonable possibility that his DNA was inadvertently transferred onto the revolver by police.

A search warrant was executed at the café by the police in 2017, which was filmed. Police found a small door leading into a wall cavity in a storage room. The cavity had space for additional storage and within it was a white pickle bucket containing three firearms and ammunition concealed in pieces of cloth, a yellow towel and a black sock. There was CCTV footage from inside the premises which showed Mr. Seifeddine accessing the wall cavity the day before the search (and the owner a week before). The camera was directed at the door leading into the cavity but did not show inside. Mr. Seifeddine was seen walking into the storage room and then into the cavity area with a white yoghurt bucket and then leaving with the yoghurt bucket after a minute or two. The defendant claimed he had tools in the bucket and was looking for an extension cord in the cavity but could not find it.

The forensic biologist at the State laboratory gave evidence regarding the defendant's DNA not being excluded as a contributor – which the Appeal Court accepted as having his DNA on five items. These were the handle of the pickle bucket, the rim of the pickle bucket, the trigger area of the revolver, the black sock the revolver was wrapped in and the yellow towel that was in the bucket. The pickle bucket handle and rim DNA may have been deposited through innocent direct transfer as 125 pickle buckets had been purchased over an 8-month period in 2017, the pickle buckets were stored in the store room, and the defendant opened them for use with the food with the rim. The Appeal Court considered the black sock and yellow towel also might have been handled by the defendant innocently.

The search video showed the police wearing gloves and visibly changing them. The lead investigator stated during the trial that he touched the door handle into the wall cavity and then touched and opened the pickle bucket lid without changing gloves. He then touched the camera when pointing out items, took the bucket by its rim and removed the bucket from the cavity by holding the handle. The court stated: 'Still wearing the same gloves he had "delved" into the bucket with both hands. He agreed that he was counting how many firearms were in the pickle bucket, using his hands to do that. He agreed that he had possibly touched every item in the bucket at that time including the towel, the sock and the "white thing" (possibly another towel) ... He then took hold of the back of a chair and dragged it over to the pickle bucket. He agreed that the pickle bucket was sitting on the chair when he removed the two magazines, still wearing the same gloves. He agreed that, in doing so, he had touched other items including the towel. Only then did he put on fresh gloves'.

'He then picked up the black sock containing the Smith & Wesson revolver and removed it from the bucket. He agreed that he had some difficulty getting the revolver out of the black sock. He was careful during that time to keep his hands off the trigger'. Footage of the search video was shown to the forensic biologist (seemingly the first time she had seen it). She accepted that there was a possibility of secondary transfer of Mr. Seifeddine's DNA from the door handle to the sock, the cloths and the firearms from the unchanged glove. (*Author note: this experiment was described (Fonnelop et al., 2015) when investigating the Meredith Kercher case.*)

Towards the end of the forensic biology evidence, the jury asked the following question: 'If a table, say, has some DNA on it, what proportion of the DNA will transfer in a touch of a latex glove? A range of values would be useful to use. Are we talking 1 to 5% or 10 to 20% or more?' The biologist was permitted overnight to produce some 'academic literature'. She found a number of studies overnight, one of which had analyzed transfer rates between plastic and cotton. The results were explained by reference to the fact that plastic is non-porous and cotton is porous. The Appeal Court stated: 'It is unfortunate that wholly new expert evidence based on a study of different materials was introduced in that way ... (the biologist) agreed that the DNA from the sock could have transferred onto the revolver at a rate of up to 76%. She agreed that the non-porous glove was put onto the sock and the porous sock was a very good receiver of DNA...also accepted that the DNA on the revolver could be explained on the basis that a third unknown individual could have handled the sock when it had the accused DNA on it and then placed the revolver into the sock'.

The Appeal Court ruled that the reasonable possibility of secondary transfer of the accused's DNA onto the firearm was not only not excluded, but indeed found some support in the evidence in the Crown case (ibid.).

7.4 The Mortuary

A mortuary may provide vast quantities of DNA due to the expulsion of body fluids during the autopsy, the transfer of DNA to multiple implements and the shedding of DNA on the autopsy table. While the autopsy may not involve the search for trace evidence as in a forensic examination, the transfer of material (whether trace or larger) from previous autopsies requires recognition and mitigation. Further, sampling of items, that may be subsequently examined for trace DNA (such as on the surfaces of the skin or internal medical samples as described in Chapter 5), requires trace DNA contamination mitigation precautions. This may indeed be difficult when dealing with vast quantities of DNA from the deceased but should be recognized as an issue, particularly in 'cold cases'.

An early investigation (Rutty et al., 2000) of 20 mortuaries in the United Kingdom was launched in the late 1990s when two instances of contamination were reported. Of those mortuaries investigated, at least half had quantifiable human DNA on instruments and mortuary surfaces. Mortuary scissors were the most frequently contaminated instrument – with profiles from one to three individuals obtained.

A later study from Germany found that DNA could be obtained from most samples taken from instruments and autopsy tables and could be linked to bodies that had been autopsied previously (Schwark et al., 2012). The study showed that in four of the six cases investigated, DNA from a previously autopsied body had been transferred via the autopsy table to the present body. The higher incidence of contamination in this study compared to the UK study was thought to result from the increased sensitivity of DNA profiling techniques in the intervening decade.

7.5 Medical Examination

The medical examination room where a complainant or suspect may be medically examined for forensic evidence requires different considerations to other medical rooms. Surfaces and implements are required to be 'DNA free' and not just 'sterile'. Equipment and instruments should be protected from the environment to ensure situations such as the Farah Jama Case (Case 1.3) do not occur again.

The ever-increasing technology that allows samples containing a few human cells, or even DNA fragments, to be analyzed requires an understanding of transfer principles in the medical examination. DNA that may be on the outside of the body may be inadvertently transferred to the inside of the body through collection tools or gloves. Case 5.2 describes conflicting results from the low vaginal and high vaginal swabs, which can only be explained by transfer of DNA through some pathway during collection.

Police officers in some jurisdictions may take samples from suspects, not only reference DNA mouth swab samples but also swabbing areas of the skin for biological material. Such procedures should also be under strict quality control.

7.6 Packaging, Handling and Storage

7.6.1 Handling

Due to direct and recurring contact with the crime scene and exhibits, there is an increased possibility for the investigator to act as a vector for the inadvertent transfer of DNA-containing material throughout the crime scene investigation and during the transportation of items to the laboratory (Fonnelop et al., 2015; Fonnelop et al., 2016).

Furthermore, improper handling of packaging during the collection and transportation phases increases the risk of contamination during the examination.

Bags containing exhibits have been assumed to be vectors of DNA transfer, yet there have been few studies in the literature until recently to understand the extent of this transfer within and outside the surface of the bags.

An early Australian study found that during transport and handling from the crime scene to the laboratory, significant quantities of DNA are frequently (a) transferred from an exhibit to the inside of its packaging and (b) transferred from its area of the initial deposit to other areas of the same exhibit and/or to other exhibits within the same package (Goray et al., 2012). Cigarette butts from two different users and un-smoked cigarettes cut to butt size were placed in A5-sized yellow paper envelopes and transported as per police conveyance to a laboratory. It was found in this study that over 85% of un-smoked butts packaged with the smoked butts had DNA profiles on them.

Another notable finding was that DNA was lost to the inside of containers holding bloodied knives. There was redistribution of DNA-containing material from the blade of the knife to other areas of the knife, including the handle.

These findings, made in 2012, highlight the need to deal with issues inherent in the collection and packaging of exhibits for forensic DNA analysis. Further study is required using the more sensitive techniques in current use today. This is important regarding the designation of a DNA profile from a 'specific' area of the item, such as the trigger on a gun, the handle of a knife or an area on a garment. The DNA may have transferred from its original area of deposition to another area during transport in the exhibit bag.

7.6.2 Packages or Exhibit Bags

The importance of the *outside* of the packaging has been recognized. A study from Norway (Fonnelop et al., 2016) demonstrated that DNA from the outside package can transfer to an exhibit within it during examination.

An Australian study showed that DNA from the exhibit could be transferred to the exterior of the exhibit bag, during the process of packing and un-packing the item (Mercer et al., 2021). A significant level of DNA could accumulate on the exterior surface of an exhibit bag through routine handling, movement or storage of an exhibit. This is unsurprising as packages are typically handled without PPE and come into contact with numerous surfaces which are not free from DNA.

More concerning is the demonstration that DNA present on the external bag surface can be subsequently transferred to the exhibit inside the bag. The Norway study (Fonnelop et al., 2016) observed one instance where a full DNA profile from an individual who handled the outside of an evidence bag was generated from a swab inside, even though it was handled far from the bag. Additional contamination events were observed, where full or partial DNA profiles from the bag handler were detected on items, when the bag was handled above the exhibit. Demonstration of indirect DNA transfer, with the exhibit bag as the intermediary vector, is of concern as it creates the potential for cross-contamination between exhibits leading to false inclusions.

A further study determined that operational forensic laboratories must consider exhibit packages as a potential source of DNA contamination and evaluate their exhibit handling and storage procedures accordingly (Mercer et al., 2022). It was stated that 'classically in forensic science we assume that a single brown paper bag is a sufficient barrier to contamination, that we can consider the inside being free of extraneous DNA and the outside being contaminated with multiple sources of DNA from its environment' (ibid.). All bags in the study were composed of brown paper, but their size and brand varied. The average DNA quantity and number of profile contributors were higher in samples taken from outside the bag before exhibit examination than after examination. Fifty-six per cent of all samples taken identified a match between DNA recovered from the evidence bag and at least one staff member. On 11 bags (out of 60), a common contributor was identified between the exhibit in the bag and the exhibit package post-examination. In one instance a DNA profile, matching that of a donor on the exhibit bag before examination, was also detected on a sample taken from the exhibit, raising the possibility of outer bag-to-exhibit DNA contamination (ibid.). A minimum likelihood ratio cut-off of 10,000 was used to minimize adventitious matches as per the particular standard laboratory protocol. Highly variable quantities of DNA were recovered from evidence bags. DNA quantities between 0.4 and 36 ng were observed pre-examination, while values between 0.07 and 37 ng were detected post-examination.

Of the 120 DNA profiles produced, 41% generated an inclusionary likelihood ratio to at least one individual on the staff elimination database, 41% did not generate any matches and 18% were too complex for analysis. At least one staff match was generated to 43% of samples taken before exhibit examination and 40% of the samples taken after exhibit examination (ibid.).

These findings reinforce the risk of bringing evidence bags, which are observed to accumulate DNA, into 'DNA-free' examination laboratories. The extensive movement and handling of exhibits, both external and internal to an operational forensic laboratory, makes it difficult and likely unrealistic to implement procedures which result in exhibit bags being free of contaminating DNA before they enter an examination laboratory (ibid.). As PPE is only required for handling exhibit packaging during the laboratory examination stage in the particular laboratory, it was expected that police and certain forensic employees may be a source of accumulating DNA. Previous studies show that DNA from individuals who have directly handled an exhibit bag without PPE can be detected on the bag's exterior (ibid.). Police and administration employees extensively handle exhibit bags during bag labelling, recording of item details and throughout the process of transporting items between storage areas. There is an opportunity for evidence bags to act as transfer vectors that facilitate the transfer of DNA from an exhibit to other exhibit bags and workspaces. Additionally, exhibit DNA may potentially accumulate within forensic workspaces, if transferred to the bag exterior during examination.

The levels of DNA detected on the exterior of evidence bags sampled within this study reinforce the risk of exhibit contamination that is posed by contact with exhibit packaging. Much of the accumulating DNA was not able to be attributed to individuals on the laboratory elimination database or the exhibit inside of the bag (ibid.).

7.6.3 Storage Facilities

Exhibits from crime scenes are stored in various types of facilities, including police laboratory storage rooms and forensic laboratory storage rooms, and even general storage areas that store documents from many years ago. Case files and documentation associated with exhibits obtained from crime scenes that are handled and stored in facilities may not necessarily be subject to strict contamination mitigation, such as within the forensic DNA laboratory.

Disturbing accounts of the state of police storage areas were described in Case 7.1. There are numerous other examples from other countries where exhibits have been lost or destroyed due to poor storage.

The following sections describe how important quality storage and packaging and handling are for an exhibit – especially when it is to be examined for trace evidence (including DNA). This impacts not only current cases but also unsolved 'cold cases', where newer more sensitive technology may be used than previously.

'Cold case' files may also contain exhibits such as hair samples and fibre samples stuck to pages of the file as part of the requirement of the particular laboratory at the time (this author information).

7.6.3.1 Storage Rooms and Shelves

Storerooms for exhibits are most often inside and have temperature-controlled environments, which may provide favourable conditions for DNA to accumulate and persist over extended durations of time. Forensic laboratories should monitor the environment or background DNA within sensitive areas such as examination benches where DNA may accumulate from the examination of multiple exhibits – hence appropriate contamination mitigation processes should be used.

However, as described above, the outside of evidence bags is recognized as efficient transfer vectors. The risk that exhibit storage locations may pose to the integrity of DNA evidence is unknown, but can be postulated as high if multiple exhibits are stored in close proximity (including touching) from multiple cases over many years, especially if there is infrequent cleaning. Shelves may contain many exhibits, which are packed together and constantly in direct contact with other exhibits, or shelves may contain few exhibits which are loosely packed and have little contact with other exhibits.

A recent study aimed to provide a better understanding of the accumulation of DNA within a forensic laboratory storage room (Mercer et al., 2023). Access to the storeroom was restricted to forensic laboratory staff and the area not routinely cleaned. Wearing of PPE is not mandatory to access the storeroom or handle exhibit packaging. Samples were collected from previously cleaned (for the experiment) forensic exhibit storeroom shelves at various time points over a 14-week period. To determine the source of accumulating DNA, profiles generated from shelf samples were compared to the laboratory staff elimination database and the DNA profiles generated from exhibits stored on each of the shelves sampled over the course of the study.

As sampling time intervals increased, there was a significant increase in DNA quantity and number of profile contributors (ibid.). The shelf height was also observed to influence the number of DNA profile contributors, with higher numbers of contributors being found on lower shelves. DNA profiles generated from the shelf samples were matched to DNA from forensic staff members who enter the storeroom and police employees, who do not enter the storeroom. There were three instances where a common DNA profile contributor was identified between a shelf sample and the profile generated from an exhibit (ibid.).

A disturbing Australian cold case murder investigation shows the problems in relying on prior procedures, especially those before the advent of DNA analysis, to try and identify an offender through DNA profiling and echoes some of the issues in Case 7.1.

CASE 7.5 MURDER OF MARIA JAMES

A divorced mother of two young sons was found dead in her home behind her suburban second-hand bookshop in 1980. She was stabbed 68 times and there are still a number of suspects (Cooper, 2022). The deputy State Coroner delivered an open finding and highlighted failures in the police investigation (English, 2022). No one has ever been charged with the murder and this inquest was the second after the initial in 1982 that ruled she had been killed by 'persons unknown'.

The body of Maria James was found in her bedroom by her ex-husband, who is not a suspect, around midday. He had been on the phone with her at his work when he heard screams and went to her home. She was found on the floor, fully clothed and with her hands tied in front. Police had at least six suspects including two local priests and a butcher who killed his ex-girlfriend in 1987. Only one of these suspects is still alive.

The coroner stated it was the mismanagement and loss of exhibits, namely Maria James' clothes and the two pillowcases, which represented the most significant lost opportunity to potentially recover DNA belonging to the assailant (English, 2022).

The crime scene photographs indicated Maria James' body and her shoes were moved whilst the photos were taken. The crime scene examiner found blood only in the bedroom and collected a quilt and blood and hair samples on one day, and then the next day the pillows, her shoes and two blood samples from the carpet. Later that same day he attended the mortuary to collect her clothing. The coroner stated: 'The pillow slips and clothes are missing and the quilt was not able to be located for arguably a decade until June 2021. In addition to the lost exhibits, a pillow from the unrelated (name) murder investigation was included with the bag containing the two pillows from the James crime scene. A DNA profile extracted from the (unrelated) pillow in 2003 was then used to incorrectly exclude persons of interest in the investigation of Mrs. James' death …. The remarkable point of the exhibit errors was that they all occurred for separate reasons and were not connected' (ibid.).

The crime scene examiner had examined the pillows and removed hairs and separated the pillow slips from the pillows, apparently to test for fingerprints in blood on the pillowslips, in 1980. There is no record of where the pillowslips went. It is possible they went for destruction along with the clothes as a biological hazard. The two pillows were subsequently in a mix up with the unrelated murder case pillow and recorded in storage as a 'quilt'.

During 2003 the forensic laboratory received 'item 9' and a forensic scientist opened it and it contained two plastic bags. The first bag was designated '9a' and was sealed and contained two pillows. The second plastic bag was unsealed and without a label, designated '9b' and contained one pillow and one pillowslip. The scientist stated they did not examine '9a' because the two pillows did not contain pillowslips and the photos from the crime scene depicted pillows and pillowslips. The scientist took samples from 9b (which were not in the crime scene photos) and identified the DNA profile of an unknown male. From 2003 the investigation proceeded on attempting to identify the unknown male from the DNA profile – believed to be the assailant – excluding persons of interest.

Another forensic scientist from the same laboratory reviewed the case in 2009 and became concerned when they could not identify the pillow and pillowcase from the crime scene photos. They e-mailed a detective in the case and suggested re-lodging the exhibits. The detective did not believe what was said raised concern about exhibit integrity.

There were two errors made regarding the quilt. The pillows were mislabelled as 'quilt' at the time of, or after, the first forensic laboratory examination for DNA in 2003, but it is unknown by whom. This error was discovered in 2010 as two exhibits had the same barcode and both were labelled 'quilt', and remedied in 2013 through a computer database entry. The second error was that the quilt itself was lost by the police, whereabouts unknown, between 2011 and 2021. A property audit was performed in February 2021 and the quilt was located.

The pillow and pillowcase from the unrelated murder were identified through another query from the forensic laboratory in 2017 as the forensic scientist reviewer could not find the pillow 9b in the crime scene photos. The reviewing detective visually matched the pillow through crime scene photos to a 1975 unsolved stabbing murder. The DNA that was used as an exclusionary tool in the Maria James investigation was found to belong to the victim in the 1975 murder.

The multiple exhibit errors in the above case that also included the mingling of exhibits from different murder scenes in the one package demonstrates the importance of the forensic scientist review in cold cases; further the importance of quality documentation, collection, handling and storage practices.

7.6.3.2 Case Files and Office Work Spaces

An Australian study on forensic case files and work spaces in forensic laboratories focused on the documents and associated work areas in which they are placed (Taylor et al., 2016). The rationale for the study was an identified contamination event that was described as *tertiary transfer* of DNA from a scientist working in an office to an item that was examined in the laboratory. The reporting scientist had not examined the item or been in the same room as the item and had no involvement in the laboratory process that yielded the contaminated DNA profile. The simplest explanation was tertiary transfer from a case file – direct contact with the case file, case file carried out to evidence recovery laboratory and reporting scientist DNA transferred to gloves of examining technician (secondary transfer) and then transferred from glove to exhibit (tertiary transfer). Case files can thus be considered as a potential DNA vector (ibid.).

Individuals who yielded the highest likelihood ratios were not necessarily the ones who most recently handled the case files in the study (ibid.). The study confirmed other research that showed the last individual who touches an item is not always going to deposit the preponderance of DNA on that item. One sample from a case file was connected to an individual who contacted a case file months prior to sampling, and other individuals who subsequently had custody of the case file were not flagged. The study also showed that items in areas that undergo high rotation cleaning will more likely give profiles with lower contributor numbers, but higher-order mixtures can also occur in these environments (ibid.).

The study above shows (1) disposable personal protection equipment should be donned prior to entering an examination area and removed on exiting and (2) case files should not be taken to examining laboratories where exhibits are being examined. This author notes that these precautionary conditions occurred in the 2000s in the United Kingdom when the author worked there.

Forensic laboratories are not the only work spaces in which exhibits are handled and associated with case files and paper work. Police laboratories and other 'screening' laboratories – indeed the crime scene itself – may have associated documents before the final forensic analysis that may retain and distribute DNA from multiple cases and people. Documentation is used at crime scenes, as well as equipment such as pens, cameras and mobile phones, as noted in Case 7.3.

The variability of deposition shown within the experiments in the study on case files (Taylor et al., 2016) was stated to add to the caution that should be used when presenting activity-level propositions to the courts. Three individuals who occupied the floor of a building produced variable results; one who appeared to deposit low levels of DNA, one who appeared to deposit high levels of DNA and a third whose DNA appeared to be deposited most in the areas they frequented the least. Any conclusions regarding the transfer of DNA from an individual onto an object brought into their regular environment would be difficult given this variability, and if the individual entered an environment for the first time it would again be difficult to determine the propensity for transfer (ibid.).

7.7 Police and Other Laboratories

There are numerous locations where exhibits may be examined prior to submission to the forensic laboratory or forensic biology laboratory. This may be at the 'scene' as described above, or at police or 'other' laboratories such as for crime scene personnel or fingerprint personnel. These laboratories may screen or sample exhibits for evidence by their own personnel (such as fingerprint examination) or for sub-sampling for DNA analysis.

7.7.1 Police and Other Laboratories Prior to Submission to Forensic Biology Laboratory

If an item is scraped, swabbed or tape lifted to obtain a sample for DNA prior to submission to the forensic DNA laboratory, then it is also necessary to ensure extraneous DNA is not introduced from surfaces or implements in those surroundings.

Environmental DNA monitoring is required to measure if cleaning procedures are efficient (Ballantyne et al., 2013). There is a need for this protocol at all locations where an exhibit is handled, such as a police laboratory or fingerprint unit, not just the DNA lab. *The necessity of police pre-examining an exhibit prior to DNA analysis should be evaluated* (Fonnelop et al., 2016; this author italics).

An Australian 2015 paper noted that in some South Australia Police examination and sampling facilities access is not restricted, the wearing of PPE is not mandatory and cleaning regimes are not regulated (Henry et al., 2015). An investigation showed that 50% of samples collected from 18 facilities had significant levels of environmental DNA. Equipment such as cameras posed the highest risk for DNA transfer with all samples yielding informative and highly mixed DNA profiles. High environmental loads were found on bench tops, cupboard/drawer handles and exhibit drying rails. This paper raises concern about awareness at police facilities where examination is performed, before submission to the forensic DNA laboratory (ibid.).

Cameras and other equipment such as crime scene kits taken to crime scenes may well be used in the crime examination laboratory or indeed at multiple crime scenes, creating a potential DNA link between different scenes if there are insufficient cleaning processes on the equipment between scenes and users. This equipment needs to be included in 'DNA monitoring' procedures.

A study from Norway (Fonnelop et al., 2016) described that the package containing an item is often opened to photograph the item and/or carry out pre-examination, such as collecting fibres and performing presumptive tests on stains, in unit areas other than the forensic biology laboratory. The regulation of the cleaning of rooms, cars and tools and wearing of protective clothing varies from unit to unit. If proper procedures are not followed, then this could be a problem for the analysis of DNA afterwards, as the original distribution of DNA could be altered.

The Norwegian study also reflected similar results to the South Australian study (Henry et al., 2015). 'Wipe tests' from 45 selected areas in two large police units were sampled for DNA quantity and quality of any resulting DNA profiles. Areas deemed *high* risk that may directly contact the exhibit had items that produced significant amounts of DNA such as a measuring tape, a ruler and a sample from a pair of gloves, with the ruler and the measuring tape producing a major DNA contributor profile that could be related to police examiners. Areas deemed *medium* risk that could be touched by gloves immediately before or after examination included a camera, a tripod, a case for DNA sampling equipment, a box with gloves, ruler, sides of roles with bench paper and a crime light that contained the most DNA. Areas deemed *low* risk that would not normally be touched during examination included a laboratory chair, a door handle and an office bench in the laboratory, which contained significant amounts of DNA. Thirteen samples contained a full DNA profile or a mixture with a major contribution, and 12 of these matched police officers working in the district.

7.7.2 Fingerprint Laboratories

An early study in 2005 found that fingerprint brushes could potentially collect and transfer DNA and the same brush could powder different items of evidence within and between crime scenes (van Oorschot et al., 2005). The dusting of latent prints may dislodge cellular debris; that debris may adhere to the brush. This brush can then potentially be used on another item where it also may transfer or dislodge cellular debris. The study recommended that when fingerprinting, the biological evidence should be avoided if possible, avoid powdering areas that may be sampled for DNA analysis, use separate fingerprint brushes and prepare and use separate aliquots of powder.

It has later been noted that secondary transfer of DNA from exhibit to exhibit can occur via fingerprint brushes and is enhanced using more sensitive DNA technologies (Bolivar et al., 2015).

Superglue chambers are also used in the fingerprint laboratory to develop latent fingerprints for easier visualization. Superglue is heated and turned to vapour in a controlled chamber with the evidence item. Because the chamber contains vaporized particles, the movement of particulates within it is possible. A study (Gibb et al., 2012) showed that DNA has the potential to accumulate and transfer within the chamber. Standards were suggested in this paper to prevent such DNA contamination.

7.8 The Forensic Biology Laboratory

7.8.1 Surfaces, Tools and Equipment

Inadvertent transfer of DNA in the laboratory may occur due to the absence of proper procedures and/or poor compliance, poor training, ineffective cleaning methods and the absence of environmental monitoring procedures (van Oorschot et al., 2019).

One way to prevent contamination is to ensure reference and evidentiary samples are kept separate in time and space and this procedure is routinely adopted in forensic laboratories; some countries even have different laboratories in different cities for separate processing. Contamination between evidentiary samples in the one case and even between cases, as seen from the above case studies, requires extra measures. There is no specific 'test' for contamination; however audit trails may follow the path of the exhibit from collection to final analysis and indicate areas in which contamination may be possible.

The laboratory environment in which the analysis is performed should be strictly controlled. Leading United Kingdom forensic laboratories had measures such as disposable laboratory overalls, hair nets, face masks and protective overshoes that were required on entry, and disposed of on exit, to the laboratory room. Weekly monitoring of the background levels of DNA in the particular laboratory room was performed to ensure contamination mitigation (author information from working in these labs in the 2000s). Exhibits from different suspects, different victims and the crime scene in the one case were examined in separate laboratory rooms, by different examiners.

All the precautions listed above are necessary for the examination of exhibits in forensic biology laboratories, where trace or large quantities of biological material may be analyzed from the same or different scenes.

The transfer of DNA from exhibit to exhibit from unrelated cases is not unknown in the forensic biology laboratory. Before the acknowledgement of the ready transfer of DNA from equipment and surfaces in a laboratory examination, where environmental DNA mitigation was not a common practice, the following case occurred.

CASE 7.6 DEATH OF JAIDYN LESKIE (JOHNSTONE, 2006)

The body of a very young child was found in a lake in Victoria, Australia, in 1998 – some 6 months after he went missing. He was dressed in a bib and trousers. A trial jury acquitted the mother's de-facto and baby sitter of murder.

A DNA profile was obtained in 2003, years after the trial, from the child's clothing. This DNA profile matched the DNA profile of a young woman who was a complainant in a rape case. Her reference DNA profile was obtained to compare to the condom in the rape case. Police could find no connection between the woman and the dead child and they lived hundreds of miles apart.

The laboratory stated there was an 'adventitious match' although there was an extremely small chance of matching the DNA from the child's clothing with a random person. The coronial inquest in 2006 discovered that the child's clothing was examined within days of the condom from the rape case, by the same forensic scientist in 1998. None of the child's own DNA was found on his clothing and the clothes had been submerged in muddy water for 6 months, which raised the question of why any DNA was found at all.

The laboratory employed a legal team to defend their position that the high probability of a DNA match was adventitious, and not contamination. The coroner noted that this meant there was a person of interest with the same DNA profile at large, which he did not accept. The coroner found that contamination had occurred in the laboratory, although the exact pathway could not be determined.

It is valuable to consider the problems arising *if* the DNA profile from the other item (any other item examined previously) had belonged to a person who may have had no alibi or corroboration of activities during the time in question.

A study from the same jurisdiction around the time of the coronial inquiry indicated the potential risk of gloves and equipment in the examination of exhibits for DNA (Poy and van Oorschot, 2006a). Gloved hands could be capable of picking up DNA-containing material from exhibits being examined, and transferred to other areas of the exhibit and/or tools whilst examining. High-risk tools such as scissor blades and forcep tongues are routinely cleaned between exhibit examinations – but their handles, containers, tissue boxes, pipettes and examination lamps that are touched by gloved hands during examination might not.

Another Victorian study (Poy and van Oorshot, 2006b) examined the levels of background DNA within a forensic laboratory. There were 195 sites examined in the laboratory and office areas and categorized according to their risk for contamination in case work. There were 52 (27%) of the areas found to have DNA fragments present. Of 32 interpretable DNA profiles there were 28 that matched staff profiles; the majority were in the office area, where no protective clothing was required. The four unknown profiles had two that matched those on the criminal database.

Any background DNA on surfaces and equipment in a laboratory examination poses a contamination risk. Environmental monitoring is one way to discover the potential background DNA present in a laboratory and to monitor cleaning processes.

In 2013 a paper from Australia (Ballantyne et al., 2013) showed that background DNA on surfaces and equipment used during forensic examination of exhibits posed a contamination risk when using the more sensitive marker systems ('Powerplex 21' compared to 'Profiler Plus' kits). Current cleaning procedures were shown to be less than adequate and it was recommended that further contamination risk issues be addressed prior to any introduction of more sensitive profiling technologies.

7.8.2 Cold Cases

Cold cases that are reinvestigated using DNA analysis on samples using the more sensitive technology require recognition that – in the past – contamination mitigation procedures may not have been suitable for such sensitive analysis.

It has been explained that 'cold cases' are an area of specific concern as exhibits may have been collected, handled and examined in conditions pre-dating the awareness of the ready transfer of DNA (Gill, 2014). As items progress from the crime scene to the DNA laboratory at each stage there is a potential for cross-contamination from other items, from the investigators themselves, and from poor packaging. However, the levels of cross-contamination can only be ascertained by proactive assessment. *There is little proactive effort to determine the extent of cross-contamination between items that are submitted to laboratories and this is clearly a weakness* (this author italics).

Previously, laboratory workers (and crime scene workers) may have worn laboratory coats or suits that were worn from week to week and not changed between cases (author information). Laboratory coats may retain DNA from previous cases/days/weeks, depending on how often they are changed and cleaned. Garments made of fabric that are not disposable, such as material laboratory coats, were worn for unknown periods of time. These coats may have deposits of DNA from the wearer, as well as deposits of DNA from items and surfaces that the clothing has contacted.

Sometimes, volunteer samples were used by laboratory scientists if they agreed, to serve as positive controls for presumptive and confirmatory serological testing (author information). For example, neat semen may have been collected from laboratory staff to use as a positive control for the screening test for seminal fluid. The disturbing case of the laboratory scientist Kevin Brown in Case 2.5 illustrates one potential consequence of contamination within the laboratory.

If exhibits are examined first in one laboratory then that laboratory needs to comply with quality assurance principles for trace DNA if the exhibits (or samples from them) are subsequently sent to another laboratory for DNA analysis. DNA environmental and contamination mitigation procedures need to be in place in all laboratories in which the exhibit occupies. This includes police laboratories or 'screening' laboratories.

A 2019 study from Australia showed that, despite strict contamination mitigation procedures in place, forensic biology examiners engaged in 'risky behaviour' with regard to gloves and touching exhibits (Goray et al., 2019). A number of laboratory-based evidence recovery personnel were videoed performing a range of examinations and several factors were assessed to evaluate the risks of DNA loss, addition and redistribution via gloves. The authors of the study observed that many different surface areas are touched by gloves during examinations, and differences exist among examiners in what they touch and when they change gloves.

7.8.3 Control Samples

Running 'negative controls' throughout the entire process is part of a DNA laboratory quality programme. There still appear to be 'blind spots' that laboratories fail to recognize may be an issue regarding contamination, even after knowledge of the sensitivity of current DNA analytical techniques. What is brought into a supposedly 'pristine' forensic DNA laboratory may contain extraneous DNA not relevant to the case, as has been demonstrated in the above sections regarding evidence bags and case files. The use of gloves by different personnel and the practice of wearing personal protection equipment may be person-dependent as also shown above. Control samples from items examined may be useful, as well as control samples from packaging and case notes.

A reagent blank control is one that consists of all reagents used in a particular method, but contains no sample. These controls do not offer protection against contamination before the process.

Elimination DNA databases of scientific staff and investigators should be maintained. If contamination is detected the contaminant DNA profile can then be searched against the elimination database.

Substrate controls have traditionally been examined to determine if the substrate (such as fabric in clothing) on which the stain was deposited, contained any material that might interfere with the interpretation of the stain. This was especially important for ABO blood grouping and enzyme typing during the 1980s (author information). In the age of DNA profiling this type of 'testing in parallel' may not be performed. It had been assumed that the 'stain' or deposit could be analyzed for DNA without reactions from any substrates on which it was deposited interfering with the test. However, this ignores pre-existing DNA on the substrate that may also be co-analyzed with the deposit, especially if techniques such as cutting out of material are used.

An early DNA case used substrate controls from a crime scene for the now notorious O.J. Simpson trial (*People v. Simpson*, 1995; Thompson, 1996). The prosecution contended that cross-contamination of blood drops at the scene of the deaths was ruled out because substrate controls (taken from unstained areas adjacent to the blood drops) were negative – that is, they contained no detectable DNA. However, it appeared that the substrate controls were not run in parallel, thus leaving open the possibility that the substrate controls were not exposed when any contamination occurred during the analysis in the laboratory.

Testing 'control' samples – from where there is no visible deposit or suspected evidential sample – from an exhibit can provide information about the presence of DNA already on that item. This is 'background DNA'. Testing areas from where the object was located (for example, the floor at a crime scene) may also give further information. If an object was picked up from the floor at a crime scene, then it may be useful to swab areas underneath and adjacent to that object. It could also be of benefit to collect a range of samples such as items in the immediate vicinity of the exhibit (van Oorschot et al., 2019).

Regarding laboratories, the only known 'laboratory' error rates are those defined where there is an outside audit of the laboratory, external proficiency tests or when there has been a corrective action documented by the laboratory. Corrective actions are those instances where samples have been known to produce a false match or other major failure. Quality assurance procedures should incorporate corrective actions. These records are often confidential. Further, it is often not communicated to outside parties whether the error is due to inadvertent transfer of DNA or some other error such as transcription error. It would be advantageous to know how often DNA transfer is involved in laboratory contamination for a specific laboratory, and even jurisdictions for crime scenes when reviewing a particular case. There are some studies in the literature which are described below.

7.9 Staff DNA Elimination Databases

7.9.1 Forensic Laboratories

An elimination database is used in forensic laboratories against which evidentiary profiles are compared, to assess if any profile on the elimination database is clearly present within the profile of the evidentiary sample (van Oorshcot et al., 2019). The elimination database should be inclusive of all who have attended a crime scene, handled an exhibit from which a DNA profile may be collected, work in areas where the exhibit is stored and/or examined, including those entering these areas for non-scientific reasons such as general cleaning, maintenance or repair (ibid.).

It is standard practice for laboratories to establish a register of staff DNA profiles to identify contamination by an operator during examination or sampling of an exhibit (Mercer et al., 2022). It is not uncommon for DNA from the persons collecting or analyzing samples to be found as contributors to DNA from the sample (Fonnelop et al., 2016).

As a result numerous international agencies including the United Kingdom (Forensic Science Regulator, 2016) recommend that personnel involved in the collection, handling and the subsequent DNA analysis of exhibits have their reference DNA samples recorded for elimination purposes.

However, this author remembers the tragic case of Kevin Brown (Case 2.5). Forensic staff members may have not only their DNA from inhabiting the forensic laboratory but also as intimate body samples such as semen used as controls for presumptive semen testing. The rationale for staff DNA databases should be made clear to other personnel in the criminal justice system.

One particular quality failure was reported at the Netherlands Forensic Institute (Kloostermann et al., 2014). An offender had been identified as 'The Avenger of Zuuk'. This case is reminiscent of the Phantom of Heilbronn.

CASE 7.8 DNA CONTAMINATION; LABORATORY STAFF

The DNA profile of an unknown woman in a series of arson and anonymous threats was identified from one of the postal items. More than 50 women volunteered to provide their DNA in a mass screening in the Dutch rural community of Zuuk.

It was later found that the supposedly incriminating DNA profile was the result of laboratory contamination. The DNA profile matched that of a Netherland Forensic Institute technician. The mass screened reference samples from the town of Zuuk were destroyed.

A survey of DNA contaminations from police services and forensic laboratories in Switzerland found that about 1% of the profiles sent to the national database were contaminated (Bassett and Castella, 2018). Reference DNA profiles of police officers collecting crime scene traces as well as forensic genetic laboratories are stored in the staff index of the national DNA database in that country in order to detect potential contamination. Within the total contamination events, 86% originated from police officers, whereas only 11% were from genetic laboratory employees and 3% were attributed to other sources (e.g., positive controls, stain–stain contaminations).

Clearly contamination mitigation procedures all along the chain of analysis of a potential DNA analysis (from crime scene to the laboratory) are necessary, to reduce the frequency of contamination incidents.

A still controversial and bizarre case from England led police to pursue what turned out to be false leads concerning the finding of a DNA profile that was believed to belong to an offender (Gill, 2014).

CASE 7.9 BODY OF THE SPY IN THE BATH

A 31-year-old Welsh mathematician and code breaker, Gareth Williams, was attached to MI6 (a UK intelligence organization) at the time of his death in 2010 (Davies, 2012). His decomposing naked body was found in the empty bathtub of his flat in London over a week after he was last seen, after concern for his welfare was expressed by his colleagues. The body was found padlocked inside a sports bag. He had no injuries, nor could illicit or poisonous substances be found in his body; the cause of death was not determined. The keys to the padlock were underneath his naked body (inside the bag), and the police originally determined he could not have locked himself inside.

A DNA profile led police to believe there was foreign DNA on the bag and body. The inquest in April 2012 heard that the transposition error of a forensic scientist when writing an e-mail asking for a DNA database check led police to believe there was foreign DNA on Williams' body. It was only discovered in February 2012 that the DNA was a partial profile belonging to a police scientist investigating the crime scene. Detectives wasted 18 months looking for a potential suspect using the DNA.

The inquest found that an unknown party had locked Gareth Williams inside his sports bag. A Metropolitan Police investigation in 2013 concluded that the death was probably accidental (BBC News, 2013).

This case shows the 'hidden perpetrator' effect (Gill, 2014) where DNA evidence was believed to belong to an offender but in fact there was no pertinent DNA evidence detected at the crime scene; the pertinent DNA belonged to a scene investigator.

7.9.2 Police Staff Contamination

The complexity and different possibilities of DNA transfer make the detection of contamination difficult, as it may result through direct or indirect transfer. It is recognized that contamination of DNA evidence by staff involved in a criminal investigation occurs predominantly through the investigating police or crime scene personnel (Bassett and Castella, 2018).

It has also been stated that without the help of databases containing reference profiles of police officers for automatic elimination, the majority of contamination incidents caused by indirect transfer would remain undetected (Neuhaber et al., 2017). This paper noted that between the years 2000 and 2016 a total of 347 contamination incidents were detected in approximately 46,000 trace samples (0.75%). Three illuminating cases were presented in the paper along with principles that apply to any crime scene investigation.

CASE 7.10 INDIRECT TRANSFER VIA CAMERA EQUIPMENT

Three years after a deprivation of liberty case there were 32 trace DNA samples submitted for analysis to the laboratory. Only one female and one male DNA profile could be detected; both DNA profiles were submitted to the national DNA database. The female DNA profile remains unknown, whereas the male DNA profile received a hit to the police elimination database. The contaminating police officer had nothing to do with the investigation, but the officer originally at the scene shared the camera equipment and this equipment was the proposed vector.

CASE 7.11 INDIRECT TRANSFER VIA CAR

A male DNA profile was identified from a hole bored in the window frame during a series of burglaries in 2015. The male DNA profile was found to be a 'hit' with the police elimination database. The police officer who caused the contamination used the car one day before the responsible crime scene investigator who drove to the crime scene.

CASE 7.12 INDIRECT TRANSFER VIA DESK

Following a burglary involving a car theft a pair of gloves was found in the stolen car. Each glove showed a different male DNA profile ('A' and 'B', respectively). As none of the profiles matched those of the suspects, they were uploaded to the national DNA database. Both subsequently matched those of police officers. The contamination of one glove (person A) was caused by the crime scene investigator who collected the evidence material. The contamination of the other glove (person B) was likely caused because the responsible crime scene investigator packed the evidence material at the desk of person B (who was not involved in the case at all).

As described in the paper (Neuhaber et al., 2017) most contamination, including the cases mentioned above, would remain undetected without the help of a database contamination system as recommended by the European Network of Forensic Sciences (ENFSI, 2016).

It was emphasized in a Norwegian study that the best contamination mitigation procedures are not always applied by police investigators (Fonnelop et al., 2016). Norway does not have a national police elimination database. The study showed 16 instances of previously undetected police staff contamination on exhibits that had been submitted between the years 2009 and 2015. Six of the cases described had police who had not been involved in the case. The conclusions were that appropriate training needed to be provided so that police are aware of the increased contamination risk with the more sensitive DNA typing systems.

References

Ballantyne, K., Poy, A., and van Oorschot, R., 2013, Environmental DNA monitoring: Beware of the transition to more sensitive typing methodologies, *Australian Journal of Forensic Sciences*, 45, 3, 323–340.

Bassett, P. and Castella, V., 2018, Lessons from a study of DNA contaminations from police services and forensic laboratories in Switzerland, *Forensic Science International: Genetics*, 33, 147–154.

Bassett, P. and Castella, V., 2019, Positive impact of DNA contamination minimization procedures taken within the laboratory, *Forensic Science International: Genetics*, 38, 232–235.

BBC News, 2013, *MI6 spy Gareth Williams death 'probably an accident' police say*, November 13.

BBC News, 2021, *Omagh bombing: Timeline of families' search for justice*, July 23.

Bolivar, P., Tracey, M., and McCord, B., 2015, Assessing the risk of secondary transfer via fingerprint brush contamination using enhanced sensitivity DNA analysis methods, *Journal of Forensic Sciences*, 61, 1, 204–211.

Cooper, A., 2022, Coroner's finding means Maria James' murder remains a mystery, *The Age*, March 31.

Davies, C., 2012, Gareth Williams inquest hears of mystery DNA at crime scene, *The Guardian*, April 24.

ENFSI, 2016, *European network of forensic sciences*, DNA Working Group, DNA database management review and recommendations, available at http://www.enfsi.eu

English, C., 2022, *Finding into death with inquest*, Coroner's Court of Victoria, Australia March 31 COR 1980 1820.

Fonnelop, A., Egeland, T., and Gill, P., 2015, Secondary and subsequent DNA transfer during criminal investigations, *Forensic Science International: Genetics*, 17, 155–162.

Fonnelop, A., Johannessen, H., Egeland, T., et al., 2016, Contamination during criminal examination: Detecting police contamination and secondary DNA transfer from evidence bags, *Forensic Science International: Genetics*, 23, 121–129.

Fonnelop, A., Ramse, M., Egeland, T., et al., 2017, The implications of shedder status and background DNA on direct and secondary transfer in an attack scenario, *Forensic Science International: Genetics*, 29, 48–60.

Forensic Science Regulator United Kingdom, 2016, Guidance 206, *The control and avoidance of contamination in crime scene examination involving DNA evidence recovery*, Issue 1.

Gibb, C., Gutowski, S., and van Oorschot, R., 2012, Assessment of the possibility of DNA accumulation and transfer in a superglue chamber: A preliminary study, *Journal of Forensic Identification*, 62, 5, 409–424.

Gill, P., 2014, *Misleading DNA evidence: Reasons for miscarriage of justice*, Academic Press Elsevier, London and New York.

Goray, M., Van Oorschot, R., and Mitchell, J., 2012, DNA transfer within forensic exhibit packaging: Potential for DNA loss and relocation, *Forensic Science International: Genetics*, 6, 158–166.

Goray, M., Pirie, E., and van Oorschot, R., 2019, DNA transfer: DNA acquired by gloves during casework examinations, *Forensic Science International: Genetics*, 38, 167–174.

Hellmann, P., *The Helmann-Zanetti Report*, On the acquittal of Amanda Knox and Raffaele Sollecito, translated into English, December 16 2011, available at www.hellmannreport.wordpress.com

Henry, J., McGowan, P., and Brown, C., 2015, A survey of environmental DNA in South Australia police facilities, *Forensic Science International: Genetics: Supplement Series*, 5, e465–e466.

Johnstone, G. 2006, *Inquest into the death of Jaidyn Raymond Leskie*. Coroner's Case 007/98 Melbourne Victoria.

Kloostermann, A., Szerps, M., and Quak, A., 2014, DNA error rates in forensic DNA analysis: Definition, numbers, impact and communication, *Forensic Science International: Genetics*, 13, 7–85.

Margiotta, G., Tasselli, G., Tommolini, F., et al., 2015, Risk of DNA transfer by gloves in forensic casework, *Forensic Science International: Genetics Supplement Series*, 5, e527–e529.

McDonald, H., 2013, Two found liable for Omagh bombing, *The Guardian*, March 7.

Mercer, C., Abarno, D., Hearnden, P., et al., 2021, DNA transfer between evidence bags: Is it a means for incidental contamination of items? *Australian Journal of Forensic Sciences*, 53, 3, 256–270.

Mercer, C., Henry, J., Taylor, D., et al., 2022, What's on the bag? The DNA composition of evidence bags pre-and post-exhibit examination, *Forensic Science International: Genetics*, 57, 102652.

Mercer, C., Taylor, D., Henry, J., et al., 2023, DNA accumulation and transfer within an operational forensic exhibit storeroom, *Forensic Science International: Genetics*, 62, 102799.

Neuhuber, F., Kreindl, G., Kastinger, T., et al., 2017, Police officer's DNA on crime scene samples – Indirect transfer as a source of contamination and its database-assisted detection in Austria, *Forensic Science International: Genetics*, 6, e608–e609.

People v Simpson 1995 Cal. Sup. Ct., LA County Case BA097211 USA.

Poy, A. and van Oorschot, R., 2006a, Beware; gloves and equipment used during the examination of exhibits are potential vectors for transfer of DNA-containing material, *International Congress Series*, 1288, 556–558.

Poy, A. and van Oorshot, R., 2006b, Trace DNA presence, origin, and transfer within a forensic biology laboratory and its potential effect on casework, *Journal of Forensic Identification*, 56, 4, 558–576.

R v Hoey [2007] The Crown Court sitting in Northern Ireland NICC 49.

Rutty, G., Watson, S., and Davison, J., 2000, DNA contamination of mortuary instruments and work surfaces: A significant problem in forensic practice? *International Journal of Legal Medicine*, 114, 56–60.

Schwark, T., Poetsch, M., Preusse-Prange, A., et al., 2012, Phantoms in the mortuary – DNA transfer during autopsies, *Forensic Science International*, 216, 1, 121–126.

Seifeddine v R [2021] NSWCCA 214 Australia.

Szkuta, B., Harvey, M., Ballantyne, K., et al., 2015, Residual DNA on examination tools following use, *Forensic Science International: Genetics Supplement Series*, 5, e495–e497.

Taylor, D., Abarno, D., Rowe, E., et al., 2016, Observations of DNA transfer within an operational forensic biology laboratory, *Forensic Science International: Genetics*, 23, 33–49.

Thompson, W., 1996, DNA evidence in the O.J. Simpson trial, *University of Colorado Law Review*, 67, 827–857.

van Oorschot, R., Treadwell, S., and Beaurepaire, J., 2005, Beware the possibility of fingerprinting techniques transferring DNA, *Journal of Forensic Sciences*, 50, 6, 1–5.

van Oorschot, R., Szkuta, B., Meakin, G., et al., 2019, DNA transfer in forensic science: A review, *Forensic Science International: Genetics*, 38, 140–166.

Vincent, Honorable F., 2010, *Inquiry into the circumstances that led to the conviction of Mr. Farah Abdulkadir Jama*, Victorian Government Printer, Melbourne.

Appendix A: Principles of Forensic DNA Transfer

The following summarizes some basic principles of the transfer of matter and the scientific method. References quoted are some core resources.

1. Principle of divisible matter

Matter divides into smaller component parts when sufficient force is applied. The component parts will acquire characteristics created by the process of division itself and retain physiochemical properties of the larger process. This principle was enunciated for the origin of evidence and the understanding of the principles of trace evidence and transfer (Inman and Rudin, 2001; Inman and Rudin, 2002).

DNA evidence originates from the human body and is thus a smaller component part that represents the larger whole in terms of discriminating one human body from another.

2. Physical transfer

Three objects and energy are required for transfer in forensic science (Inman and Rudin, 2001).

Object 1 – The original source.
Object 2 – A fragment (ultimately the evidence) derived from the source.
Object 3 – A target object to which the fragment is transferred.
Energy – Facilitates movement of the fragment from the source to the target. This may be contact between the source and the target, but it need not necessarily be so.

3. Scientific method and hypothesis testing

The basis of all forensic science work uses the scientific method. It is a method of study used to try and understand the physical universe. The scientist obtains qualitative data by observation and quantitative data by measurement.

Classical science is defined by the notion of hypothesis testing. First a hypothesis, or theory, is proposed. Experiments are then performed to test this hypothesis. The results of the experiments will support, refute or be equivocal regarding the hypothesis. The scientific method provides the framework for testing of the hypothesis.

Consideration of more than one hypothesis to account for the physical evidence forces the examiner to think critically and discourages bias in evaluating the results. Alternate hypotheses should be formulated before analysis of the results and the examiner should not feel restricted to just one set of alternative hypotheses – the evaluation may demand more. Indeed, further information may demand a re-evaluation and further hypothesis testing.

Hypotheses should be articulated clearly and transparent to the reader so that the justice system understands what is being evaluated. The examiner may be at odds with what is required and unless hypotheses are understood and reasonable the analysis may be useless (Inman and Rudin, 2001; Gill, 2016).

4. Statement of limitations

The statement of limitations is the starting point for *trace* DNA evidence reports where there are only DNA profiling results (Giil, 2014); that is, a sub-source evaluation that is useful for *discrimination* between people, not *how* the DNA got to where it was recovered (or how it was transferred).

The limitations demonstrate to the reader the type of evidence that is being evaluated and the commensurate uncertainty. It should be in the body of the report and not tucked away in some appendix where the reader may not readily associate the section to the report.

- Although a DNA profile has been obtained, it is not possible to identify the type of cells from which the DNA originated nor to state when the cells were deposited.
- It is not possible to make any conclusion about transfer and persistence of DNA.
- Because the DNA test is very sensitive it is not unusual to find mixture DNA profiles. If the potential origin of DNA profiles cannot be identified, this does not mean this is relevant to the case due to the transfer of cells just through casual contact.

Variations of the above can be produced according to the DNA evidence evaluated but the 'starting' point is useful to remember.

The forensic scientist should provide the court with an unbiased list of all possible modes of DNA transfer (Gill, 2014).

5. Framework of propositions

Interpretation of evidence within a 'framework of propositions' describes the various levels at which evidence may be evaluated (Cook et al., 1998; Gill, 2014; Gill et al., 2020).

The framework provides a 'hierarchy' where the probative value of DNA evidence increases at each level.

- The sub-source level refers to an evaluation of the DNA profile.

- The source level refers to an evaluation of the DNA profile that can be associated with a specific body fluid such as blood or semen.
- The activity-level associates the DNA profile with an activity, e.g., sexual assault.
- The highest level is the ultimate issue of guilt/innocence.

The court deals with the highest level of the ultimate issue; this is not within the remit of a scientist.

6. 'Takeaways' regarding DNA transfer

The National Institute of Standards (USA) provided a summary of our knowledge to date regarding core concepts of trace DNA transfer in a draft in the document on DNA mixtures (NIST, 2021).

Chapter 5 described some 'takeaways' relating to the study of DNA transfer and listed some accepted basics regarding DNA transfer.

#5.1: DNA can be transferred from one surface or person to another, and this can potentially happen multiple times. Therefore, the DNA present on an evidence item may be unrelated (irrelevant) to the crime being investigated.

#5.2: Highly sensitive DNA methods increase the likelihood of detecting irrelevant DNA. When assessing evidence that involves very small quantities of DNA, it is especially important to consider relevance.

#5.3: Highly sensitive methods increase the likelihood of detecting contaminating DNA that might affect an investigation. Contamination avoidance procedures should be robust both at the crime scene and in the laboratory.

#5.4: DNA statistical results such as a sub-source likelihood ratio do not provide information about how or when DNA was transferred, or whether it is relevant to a case. Therefore, using the likelihood ratio as a standalone number without context can be misleading.

#5.5: The fact that DNA transfers easily between objects does not negate the value of DNA evidence. However, the value of DNA evidence depends on the circumstances of the case.

#5.6: There is a growing body of knowledge about DNA transfer and persistence, but significant knowledge gaps remain.

References

Cook, R., Evett, I., Jackson, G., et al., 1998, A hierarchy of propositions: Deciding which level to address in casework, *Science and Justice*, 38, 231–239.

Gill, P., 2014, *Misleading DNA evidence: Reasons for miscarriage of justice*, Academic Press Elsevier, London and New York.

Gill, P., 2016, Analysis and implications of the miscarriages of justice of Amanda Knox and Raffaele Sollecito, *Forensic Science International: Genetics*, 23, 9–18.

Gill, P., Hicks, T., Butler, J., et al., 2020, DNA commission of the international society for forensic genetics: Assessing the value of forensic biological evidence - Guidelines highlighting the importance of propositions. Part II: Evaluation of biological traces considering activity level propositions, *Forensic Science International: Genetics*, 44, 1–13.

Inman, K. and Rudin, N., 2001, *Principles and practice of criminalistics: The profession of forensic science*, CRC Press, Boca Raton, Florida.

Inman, K. and Rudin, N., 2002, The origin of evidence, *Forensic Science International*, 126, 11–16.

NIST, National Institute of Standards and Technology, 2021, *DNA Mixture Interpretation, A NIST Scientific Foundation Review*, NISTIR 8351-DRAFT, June.

Appendix B: Flawed Logic and Forensic DNA Transfer

Interpreting forensic DNA evidence, particularly trace DNA evidence and concepts of transfer, may be susceptible to flawed logic and reasoning by the analyst. It is necessary that the following issues are recognized.

A. Effects

There are particular effects that may produce miscarriages of justice and/or flawed thinking/logic in forensic DNA cases and may be relevant to forensic DNA transfer.

1. Naive Investigator Effect

The tendency for the DNA evidence to override any neutral or exculpatory evidence is called the 'naive investigator effect'.

As an example, the 'match' between DNA profiles (such as from a vaginal swab and a reference sample) occurs because of a contamination incident in collection, handling and/or forensic examination.

The 'naive investigator' finds the closest match to a crime stain using a DNA database, ignores exculpatory evidence and seeks to prosecute the matching individual. Further, there does not have to be a DNA database search, any DNA profile found at a crime scene will do (Gill, 2014).

The naive investigator effect was inspired by the Adam Scott Case (see Case 5.1). It can also be seen in the case of Farah Jama (Case 1.3).

2. Hidden Perpetrator Effect

A dangerous effect of DNA contamination occurs when insufficient DNA is recovered from an actual perpetrator (Gill, 2014). The contaminant may be the only visible DNA profile. Using an example of sexual assault there is an expectation driven by confirmation bias that the DNA has come from the culprit. If the database trawl identifies an individual they now become the only (innocent) suspect.

Donors of background and investigator-mediated contaminant DNA will automatically become suspects if the perpetrator DNA is absent from the DNA profile.

This effect is demonstrated in the case of the MI6 agent Gareth Williams (see Case 7.9).

3. Compounded Error Effect

The 'compounded error effect' combines the three effects of the association fallacy (below), the hidden perpetrator effect and the naive investigator effect. False deductive bias driven by confirmation bias leads to a cascade effect that propagates multiple errors that address the ultimate issue of guilt/innocence (Gill, 2014).

This effect is demonstrated by the case of Adam Scott (see Case 5.1).

The compounded error effect is similar to the definition of 'bias cascade" (see below).

4. The Serial Error Effect

The same kinds of errors are propagated across unrelated offences. The Phantom of Heilbronn (Case 3.1) and the Avenger of Zuuk (Case 7.8) are examples of the 'serial error' effect.

5. The Swamping Effect

The effect occurs when there is a very large likelihood ratio for the DNA profile observed in favour of the suspect/accused contributing to the profile examined – compared to not contributing (such as one billion).

This large number overrides any other information that may suggest the defendant is innocent. A court may not be provided with an indication of the error rate in the laboratory or provided with the information that the suspect was discovered through a database trawl. The court may be assured by the scientist that no contamination has occurred and this is extended to the case as a whole (Gill, 2014).

This effect is illustrated by the case of Farah Jama (see Case 1.3).

B. Fallacies

Fallacies are errors in thinking that are perpetuated over many cases such that they become entrenched in the criminal justice system. The below fallacies are encountered in cases that incorporate principles of forensic DNA transfer.

1. Transposition of the Conditional

Many types of fallacies in forensic science evaluation incorporate transposition of the conditional – transposing the probability of the evidence observed given a scenario to the probability of the scenario given the evidence observed. This occurs particularly when using the likelihood ratio to evaluate forensic DNA results in the meaning of the case. It is important to remember *the scientist evaluates the evidence observed*, they cannot do more.

Sometimes this fallacy is called the 'prosecution fallacy'. However, the fallacy is committed by many and is not confined to a prosecutor or the prosecution in general.

The literature has tried to illustrate the problem but unfortunately has not been successful as it is still repeated in trial testimony, court rulings and Appeal Court rulings so that the fallacy becomes entrenched.

An oft-quoted illustration of the basic fallacy follows (Aitken et al., 2010):

1. **If** I am a monkey, I have two arms and legs.
2. **If** I have two arms and legs, I am a monkey.

The fallacy is obvious from the above transposition, and conditions. Clearly, Proposition 1 is true (unless the monkey is injured) whereas Proposition 2 is false.

Inserting probabilities this becomes:

- What is the probability I have two arms and legs, **if** I am a monkey? (Hopefully '1').
- What is the probability I am a monkey, **if** I have two arms and legs? (Clearly a fraction considering 6 billion humans).

The above is a problem of conditional probabilities upon which likelihood ratios are dependent. Because likelihood ratios are the only suitable method to interpret DNA profiles that are low level, poor quality or have the potential for 'drop out' of a fragment so that the donor may not be fully represented, it is necessary to understand the fallacy if likelihood ratios are required.

When using the likelihood ratio, the probability of the evidence observed is viewed in light of various alternative scenarios – that is, the probability of the evidence is conditioned on a scenario, assuming it to be true. It does not mean it is true. Varying scenarios may be proposed. Further, the defence does not have to accept what the prosecution proposes for the defence scenario as it is their domain.

Thus for the prosecution scenario the probability of the evidence observed given the prosecution scenario is true is Pr E/Hp, and is the numerator of the likelihood ratio. The probability of the evidence observed given the scenario of the defence is true is PrE/Hd and is the denominator of the likelihood ratio. This likelihood ratio (because it is a ratio of probabilities) is PrE/Hp divided by PrE/Hd.

The transposition of the conditional occurs when the probability of the evidence observed is then transposed to become the probability of the scenario – sometimes the probability of the scenario given the evidence, where the transposition is clearer to a reader. It is not the remit of the scientist to determine the probability of a scenario and indeed they cannot do so. The scientist can only evaluate the evidence observed.

2. The Association Fallacy

The strength of the evidence of the DNA profile may be wrongly transposed to include a defined body fluid such as blood or semen in the calculation, or indeed an 'activity'. A probability is wrongly transposed (upwards) from one level of the framework of propositions to another (Gill, 2014; Gill et al., 2020).

The 'association fallacy' describes the automatic assumption that the DNA profile has come from a body fluid based on the strength of presumptive or RNA tests, verified by 'expert opinion'.

The observation of a body fluid and the detection of a DNA profile are two separate tests.

The fallacy also describes the wrongful association of the presence of a body fluid, such as blood or sperm in semen, with an activity such as violence or sexual assault.

This fallacy assumes that there is a *dependency* between two observations or events. This dependency cannot be assumed unless there is experimental evidence in support.

3. Source Level Fallacy

The source level fallacy occurs when a sub-source likelihood ratio (for a DNA profile) is incorrectly transposed to the source level. As an example, a likelihood ratio for a DNA profile may be one billion in favour of the suspect contributing to the profile at the sub-source level, but this is then transposed to one billion in favour of the suspect contributing to the blood, if blood has been assumed as 'the source'.

Testing for DNA and testing for blood, semen or saliva are separate tests.

An example of the fallacy is the Adam Scott Case (see Case 5.1). The DNA was assumed to have come from sperm but in fact came from saliva in an unrelated matter.

Assumptions are that the DNA comes from just one individual, that the specified body fluid is present in the sample and the probability of observing this is one. Another assumption is that this is the only material in the sample producing DNA.

If a DNA profile, especially a low-level DNA profile, is obtained from a dilute or mixed body fluid then the biological nature of the DNA profile may be disputed and therefore the sub-source level cannot be elevated to the source level (Gill et al., 2018). Similarly if a mixed DNA profile is obtained, even from a confirmed body fluid, it may be unclear as to which DNA profile can be attributed to the body fluid.

Control samples of adjacent material may be useful to ascertain the level of 'background DNA' or 'prevalent DNA' that may be in the sample.

4. Activity-level Fallacy

The activity-level fallacy occurs when either the sub-source level (DNA profile) or the source level (DNA profile of a specific biological fluid) likelihood ratio is transferred to activity scenarios, specifically when the scientist uses the phrase 'in my opinion'. An example is that of Adam Scott (Case 5.1) where the sub-source level was transferred to the source level and then the activity level. Direct transfer was assumed but indirect transfer through contamination was the reality.

The evaluation of DNA evidence given activity-level propositions requires the scientist to consider factors associated with the transfer, persistence, prevalence and recovery of DNA (DNA-TPPR). Further, there should be a consideration of error in the context of the case. The primary source of information should be empirical data where the 'ground truth' is known rather than 'casework experience'.

The activity-level fallacy may occur when the scientist is asked in court about 'activity level' propositions and the only proposition that they have considered is the sub-source level of the DNA profile. Providing opinions on activity-level propositions require extra time, data and consideration (Meakin et al., 2021).

5. The Prosecutor's Fallacy

This fallacy occurs when there is a 'transposition of the conditional'. Instead of stating the probability of the evidence in light of (or conditioned on) two alternative scenarios or hypotheses, the fallacy occurs when it is stated as the probability of that scenario or a ratio of probabilities of alternative scenarios.

As an example of the fallacy, in its most basic form at the sub-source level for a three-person mixture, the probability of observing the DNA profile conditioned on the suspect and two unknowns as contributors versus three unknowns as contributors is the likelihood ratio. This becomes abbreviated in testimony or reference to such as the probability of the suspect and two others versus the probability of three unknowns as contributors.

The scientist can only evaluate the evidence, not the scenarios. The 'conditioning' aspect becomes lost in the translation.

C. Misconceptions in Forensic DNA Transfer

1. DNA believed to be deposited through 'touch' is from shed skin cells.

DNA transferred from the surfaces of the skin, such as the skin on the hands used in 'touching' or 'handling' an item had previously, and may still, stated to be from shed skin cells. The terms 'touch DNA' and 'contact DNA' further confuse the issue (see Chapter 3).

Sources of DNA from skin may be from other areas of the body (such as nasal secretions or saliva) and this may be undetected due to the absence of testing for other body fluids particularly due to insufficient sample. Extracellular DNA conveyed by body secretions has been shown to be a considerable portion of the amount of DNA shed from the skin (see Chapter 4).

2. DNA from the last handler, user or wearer will be detected as the major contributor.

Persistence of the original user and detection of subsequent users is dependent on a multitude of factors such as the type of substrate material, the amount deposited, the sampling methods, detection tools and interpretation methods.

It is not possible to determine 'who' made the last deposit of DNA, or who last 'handled' an item.

D. Cognitive Bias

1. Confirmation Bias

Confirmation bias refers to the tendency to search for and interpret information that confirms our prior beliefs.

The investigator or scientist searches for evidence to discover a perpetrator of some offence. There is an expectation that a DNA profile must be significant in relation to a crime due to Locard's Exchange Principle that every contact leaves a trace. The issue is the relevance of that DNA profile to the case.

Using the scientific method the scientist considers their findings using alternative hypotheses and does not become fixated on one and search for support for that hypothesis in light of their findings.

2. Cognitive Dissonance

Investigators/scientists may discredit, ignore or explain away information that does not support their views.

This bias is exemplified by the case of Kevin Brown (see Case 2.5).

3. Bias Cascade and Bias Snowball (Dror, 2018)

'Bias cascade' is where bias cascades from one component of an investigation to another. Countering bias cascade may be achieved by controlling the information flow between the different stages of the investigation.

'Bias snowball' has the bias increasing in strength and momentum as different components of an investigation influence one another. Bias is not only cascading from one stage to another, but bias increases as irrelevant information from a variety of sources is integrated and influences each other. As one piece of evidence influences another, then greater distortive power is created because more evidence is affected (and affecting) other lines of evidence, causing bias with greater momentum, resulting in the increasing snowball of bias (Dror, 2018).

Bias cascade and bias snowball can be minimized by 'compartmentalization' of examiners and evidence.

References

Aitken, C., Roberts, P., and Jackson, G., 2010, Practitioner Guide 1: Fundamentals of probability and statistical evidence in criminal proceedings, London Royal Statistical Society (available online).

Dror, I., 2018, Biases in forensic experts, *Science*, 360 (6386), 243.

Gill, P., 2014, *Misleading DNA Evidence: Reasons for miscarriage of justice*, Academic Press Elsevier, London and New York.

Gill, P., Hicks, T., Butler, J., et al., 2018, DNA commission of the international society for forensic genetics: Assessing the value of forensic biological evidence – Guidelines highlighting the importance of propositions, Part I: evaluation of DNA profiling comparisons given (sub)-source propositions, *Forensic Science International: Genetics*, 36, 189–202.

Gill, P., Hicks, T., Butler, J., et al., 2020, DNA commission of the International society for forensic genetics: Assessing the value of forensic biological evidence - Guidelines highlighting the importance of propositions. Part II: Evaluation of biological traces considering activity level propositions, *Forensic Science International: Genetics*, 44, 1–13.

Meakin, G., Kokshoorn, B., van Oorschot, R., et al., 2021, Evaluating forensic DNA evidence: Connecting the dots, *WIREs Forensic Science*, 3(4), July/August, e1404.

Glossary of Terminology

Activity level: evaluating the analytical findings of the trace material with competing scenarios as to activities, such as how the material got to its location; a proposition relating to an activity that is relevant to the deliberations of the court

Aerosol transfer:: transfer of DNA achieved without an intermediary by shouting, speaking, spitting or coughing DNA into the environment

Allele: variation of the same sequence of short tandem repeats at a specific place on a chromosome; in numeric form; humans inherit two copies, one from each parent; may be the same (homozygote) or different (heterozygote)

Artefact: a result occurring in the DNA profile as a result of the process rather than intrinsic to the DNA.

Background DNA: DNA on a surface that has been acquired through previous use or already present on a surface before the crime event; from one or more unknown individuals

Bi-directional transfer: exchange of material when two surfaces come into contact

Cold cases: criminal cases that have been unsolved; cases where alleged perpetrator has not yet been identified

Confirmatory test: used to confirm the presence of a particular biological material such as blood or semen; minimal false positives

Contact DNA: proposed to have been deposited through contact but not known unless in controlled experimental studies; not appropriate term

Contamination: material deposited after the crime event such as through collection, handling or examination; the transfer of irrelevant DNA during an investigation; inadvertent introduction of DNA into a DNA sample

Degradation: the breakdown of the DNA strand through age, environment, bacterial or chemical insult resulting in a greater loss of the longer fragments compared to the shorter fragments

Differential extraction: separation of sperm from the other material in semen; performed in the laboratory, using different extraction solutions and centrifugation; aims to produce two separate DNA profiles – one from the first fraction with less harsh solvent and one from the second, denser, fraction with solvent designed to rupture sperm; may be incomplete with carry-over from one fraction to another

Differential sampling: rationale for sampling by using different and the most appropriate technique for the biological matter or other trace material in the deposit area

Direct or primary transfer: transfer of material from the source to another object; first transfer step

DNA: deoxyribonucleic acid; a complex molecule with a genetic code; the building block of life

DNA profile: a graphical and numerical representation of testing at various genetic sites on the DNA molecule

DNA-TPPR: DNA transfer, persistence, prevalence and recovery

Elimination samples: reference samples from persons used for DNA profile comparison purposes; believed not to be a suspect but may have left DNA behind at scene at some time; could include investigators, family members of suspect, co-workers of suspect; should not be stored on databases unless forensic personnel with their consent

Elimination databases: DNA profiles from personnel involved in collection, handling and processing of exhibits subsequently analyzed for DNA; DNA profiles stored on a database, often internally in a laboratory

Epithelial cells: cells lining the organs of the body; includes the skin and the vagina cavity

Extracellular DNA: eDNA; DNA not purported to be from a cell in the deposit

Forensic analysis: scientific tests or techniques relevant to legal proceedings

Framework of circumstances: background information that summarizes all of the circumstances relevant; provisional in nature, it might be changed by new information; used in formulation of hypotheses

Gene: a site on the DNA molecule; sequence of DNA base pairs

Ground truth: information provided by direct observation (i.e., empirical evidence) as opposed to information provided by inference; a situation where the correct answer is known by design

Handler DNA: DNA believed to be from 'touching' or 'handling' an item; inappropriate term as cannot determine DNA transferred by the hand; 'trace DNA' appropriate

Hierarchy of propositions: propositions or hypotheses that address issues of interest to the court; propositions are only addressed in pairs, at different levels (sub-source to offence level); there will be two propositions that respectively represent the prosecution (Hp) and the defence (Hd) positions relating to the issue

Investigator-mediated contamination: transfer of DNA via an investigator's person or equipment; may not be detected or overlooked; tertiary or higher transfer

Indirect transfer: transfer when the material deposited from the source on a surface is then transferred to another surface. There has been no physical contact between the original depositor and the final surface on which the deposit is located. Further transfer may follow involving other intermediaries.

Likelihood: conditional probability of an event, where the event is considered as an outcome corresponding to one of several conditions or hypotheses

Likelihood ratio (LR): the probability of the evidence under one proposition divided by the probability of the evidence under an alternative, mutually exclusive proposition

Mixed DNA: a sample that contains DNA from two or more people

Multi-step transfer: a number of transfers from item to item or within that item

ng: nanogram; a billionth of a gram (10 to the power of –9 g); there is 1 ng of DNA in ≈150 human cells

Non-self DNA: DNA not from the individual purported to have the DNA on them; DNA not from the donor to an object or deposit

Offence level: a proposition that relates to a criminal offence, e.g., sexual assault that is relevant to the deliberations of the court; domain of the court and not the scientist

Persistence of DNA: a variable of survival of the DNA in a deposit; depends on other variables such as environmental conditions; difficult to predict regarding time as other variables to consider; part of the understanding of DNA-TPPR

Person of interest (POI): also referred to as a significant individual; a person whose DNA profile is the subject of the evaluation

pg: picogram; a trillionth of a gram (10 to the power of –12 g); there are ≈6 pg of DNA in a single diploid human cell

PPE: personal protection equipment; disposable; changed between cases and sometimes within cases to prevent inadvertent DNA transfer; designed to protect the individual from the environment, and the environment from the individual; typically gloves, mask, hair net, disposable suits/overalls, overshoes

Prevalent DNA: DNA from a known individual in the context of the case; deposition unrelated to the event under consideration

Prevalence of DNA: part of the understanding of DNA-TPPR; used in activity level propositions

Presumptive tests: a screening test used to indicate the possible presence of the named body fluid

Probability: a number between zero and one that reflects in a reasonable way our belief that the event is true

Propositions: the hypothesis of the defence or prosecution arguments that are used to formulate the likelihood ratio; should be stated clearly

Quaternary transfer: third indirect transfer taking place after tertiary transfer; fourth transfer step

Recovery of DNA: obtaining DNA containing material from an item; may be different methods available for the one deposit; dependant on substrate and deposit material

Scientific method: a systematic method for acquiring knowledge about the world around us; characterized by formulation of hypotheses and testing them by experiments and gathering empirical data from observations

Secondary transfer: first indirect transfer of material after primary or direct transfer, second transfer step; taking place via an intermediary object or surface

Self DNA: DNA from the individual concerned; DNA from the person considered to have directly deposited DNA from their person

Shedder status: the propensity of individuals to leave DNA behind when contacting a surface with the skin; now believed to be on a 'continuum' without being a fixed status

Stutter: an artefact caused by 'slippage' in the copying process in the amplification step of DNA analysis

Source level: a proposition relating to the origin of a DNA profile that has been attributed to a body fluid or tissue

Stochastic: random variation

Sub-source level: a proposition that relates only to the DNA profile; based solely on discriminating power

Tertiary transfer: second indirect transfer of material after secondary transfer; taking place via an intermediary object/s or surface/s; third transfer step

Touch DNA: commonly used to describe DNA collected from surfaces believed to have been 'touched' by a hand; now accepted as not appropriate as currently cannot determine if surfaces have been touched (sometimes referred to as 'skin DNA' or 'contact DNA')

Trace DNA: DNA that cannot be attributed to a body fluid; sometimes referred to trace quantity such as below recognized thresholds

Transfer: movement of matter from one place to another; characterized by the source, the recipient and the environment

Wearer DNA: DNA that is associated to habitual wearer of garment

Index

E

F

G

H

I

V

W